SpringerBriefs in Applied Sciences and Technology

Computational Intelligence

Series Editor

Janusz Kacprzyk, Systems Research Institute, Polish Academy of Sciences,
Warsaw, Poland

SpringerBriefs in Computational Intelligence are a series of slim high-quality publications encompassing the entire spectrum of Computational Intelligence. Featuring compact volumes of 50 to 125 pages (approximately 20,000-45,000 words), Briefs are shorter than a conventional book but longer than a journal article. Thus Briefs serve as timely, concise tools for students, researchers, and professionals.

Svetlin Georgiev

Neural Network Methods for Dynamic Equations on Time Scales

 Springer

Svetlin Georgiev
Department of Mathematics
Sorbonne University
Paris, France

ISSN 2191-530X   ISSN 2191-5318   (electronic)
SpringerBriefs in Applied Sciences and Technology
ISSN 2625-3704   ISSN 2625-3712   (electronic)
SpringerBriefs in Computational Intelligence
ISBN 978-3-031-85055-4   ISBN 978-3-031-85056-1   (eBook)
https://doi.org/10.1007/978-3-031-85056-1

Preface

Time scale theory was first initiated by Stefan Hilger in 1988 in his Ph.D. thesis to unify both approaches of dynamic modeling: difference and differential equations. Similar ideas have been used before and go back in the introduction of the Riemann-Stieltjes integral which unifies sums and integrals. Many results to differential equations carry over easily to corresponding results for different equations, while other results seem to be totally different in nature. Because of these reasons, the theory of dynamic equations is an active area of research. The time scale calculus can be applied to any field in which dynamic processes are described by discrete or continuous time models. So, the calculus of time scales has various applications involving non-continuous domains such as certain bug populations, phytoremediation of metals, wound healing, maximization problems in economics, and traffic problems.

In recent decades, among the various machine intelligence procedures, artificial neural network (ANN) methods have been established as powerful techniques to solve a variety of real-world problems because of the ANN excellent learning capacity. ANN is a computational model or an information processing paradigm inspired by the biological nervous system.

The main aim of this book is to handle dynamic equations on time scales using ANN. The book contains five chapters. In Chap. 1, basic facts and methods for ANN modeling are considered. In Chap. 2, the multilayer artificial neural network (ANN) model is introduced for solving of dynamic equations on arbitrary time scales. A multilayer ANN model with one input layer containing a single node, a hidden layer with m nodes, and one output node are considered. The ANN trial solution of the considered dynamic equations is represented as a sum of two terms. The first term satisfies the initial or boundary conditions. The second term contains the output of the ANN with adjustable parameters. The feed-forward neural network model and unsupervised error back-propagation algorithm are investigated. Modification of network parameters is done without the use of any optimization technique. In Chap. 3, the regression-based neural network (RBNN) model is introduced for solving dynamic equations on arbitrary time scales. The RBNN trial solution of dynamic equations is obtained by using the RBNN model for single input and single output system.

The number of nodes in the hidden layer has been fixed according to the degree of the polynomial in regression fitting, and the coefficients involved are taken as initial weights to start with neural training. In this chapter, a variety of initial and boundary value problems are solved. In Chap. 4, the Chebyshev neural network (ChNN) model is developed for solving dynamic equations on arbitrary time scales. The ChNN trial solution of dynamic equations is obtained by using the ChNN model for single input and single output system. Initial value problems and boundary value problems are solved. Chapter 5 is devoted on the Legendre neural network modeling for solving initial and boundary value problems for dynamic equations on time scales.

This book is addressed to a wide audience of specialists such as mathematicians, physicists, engineers, and biologists. It can be used as a textbook at the graduate level and as a reference book for several disciplines.

Paris, France   Svetlin Georgiev
September 2024

Contents

Chapter 1
Introduction

Artificial neural network (ANN) is one of the areas of Artificial intelligence (AI) research. It is also an abstract computational model based on the organizational structure of the human brain. ANN is a data modeling tool that depends on various parameters and learning methods. Neural networks are typically organized in layers that are made up of a number of interconnected "neurons/nodes" which contain "activation functions".

The ANN method has been established as a powerful technique to solve a variety of real-world problems. This method has been successfully applied in various fields such as function approximation, clustering, prediction, identification, pattern recognition, solving ordinary and partial differential equations.

1.1 Architecture of ANN

This is a technique that seeks to build an intelligent program using models that simulate the working of neurons in the human brain. The key element of the network is the structure of the information processing system. The network is composed of a large number of highly interconnected processing elements (neurons) working in parallel to solve specific problem.

Neural computing is made up of a large number of Artificial neurons and a larger number of interconnections among them. According to the structure of these interconnections, there are different classes of neural network architecture

In a feed-forward neural network, neurons are organized in the form of layers. Neurons in a layer receive input from the previous layer and feed their output to the next layer. Network connections to the same or previous layers are not allowed. The data goes from the input node to the output node in a strictly feed-forward way. The output of any layer does not affect the same layer. Let

© The Author(s), under exclusive license to Springer Nature Switzerland AG 2025
S. Georgiev, *Neural Network Methods for Dynamic Equations on Time Scales*,
SpringerBriefs in Computational Intelligence,
https://doi.org/10.1007/978-3-031-85056-1_1

$$X = \begin{pmatrix} x_1 \\ \cdots \\ x_n \end{pmatrix} \quad \text{and} \quad O = \begin{pmatrix} o_1 \\ \cdots \\ o_n \end{pmatrix}$$

be the input vector and the output vector, respectively, and f be the activation function. Let also, $W = (w_{ij})$,

$$W = \begin{pmatrix} w_{11} & \cdots & w_{1n} \\ \cdots & \cdots & \cdots \\ w_{n1} & \cdots & w_{nn} \end{pmatrix}$$

be the weight matrix or connection matrix. Then

$$WX = \begin{pmatrix} w_{11} & \cdots & w_{1n} \\ \cdots & \cdots & \cdots \\ w_{n1} & \cdots & w_{nn} \end{pmatrix} \begin{pmatrix} x_1 \\ \cdots \\ x_n \end{pmatrix}$$

$$= \begin{pmatrix} \sum_{j=1}^{n} w_{1j} x_j \\ \cdots \\ \sum_{j=1}^{n} w_{nj} x_j \end{pmatrix}$$

is the net input value. In Fig. 1.1, it is shown a feed-forward ANN. In a feedback neural network, the output of one layer routes back to the previous layer. This network can have signals traveling in both directions by the introduction of loops in the network. This network is very powerful and gets extremely complicated. All possible connections between neurons are allowed. Obviously, feedback neural networks are used in optimization problems. In Fig. 1.2, it is shown a diagram of a feedback neural network.

1.2   Learning Processes

Learning is one of the characteristics of the ANN model. Any neural network processes knowledge that is contained in the values of the connected weights. Most

Fig. 1.1 A feed-forward neural network

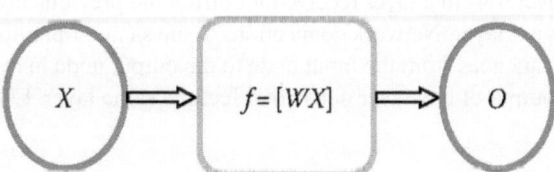

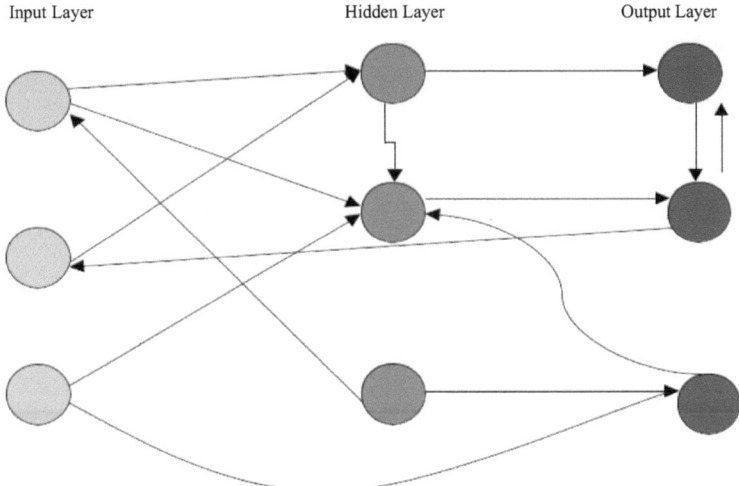

Fig. 1.2   A feedback neural network

ANNs contain some form of "learning rule" that modifies the weights of the connections according to the input patterns that is presented with. There are various kinds of learning rules used by neural networks such as: error back-propagation learning algorithm (delta learning rule), hebbian learning rule, perceptron learning rule, widrow-hoff learning rule and winner-take-all learning rule, and so on.

The error back-propagation learning algorithm is also known as the delta learning rule and it is one of the most commonly used learning rules. A simple perceptron can handle linearly separable or linearly independent problems. Using the partial derivative of error of the network with respect to each of its weights, we can know the flow of error direction of the network. If we take negative derivative and then proceed to add it to the weights, the error will decrease until it approaches a local minimum. We have to add a negative value to the weight or the reverse if the derivative is negative. Then, the partial derivatives are applied to each of the weights, starting from the output layer weights to the hidden layer weights, and then from the hidden layer weights to the input layer weights.

In general, the training of the network involves feeding samples as input vectors, calculating the error of the output layer and then adjusting the weights of the network to minimize the error. The average of all square errors E for the outputs is computed to make the derivative simpler. After the error is computed, the weights can be updated one by one. The descent depends on the gradient ∇E for the training of the network.

Let us consider a multilayer neural architecture containing one input node x, three nodes in the hidden layer y_1, y_2, y_3, and the output node O (see Fig. 1.3). Now, by applying feed-forward recall with error back propagation learning to the model given in Fig. 1.3, we get the following algorithm.

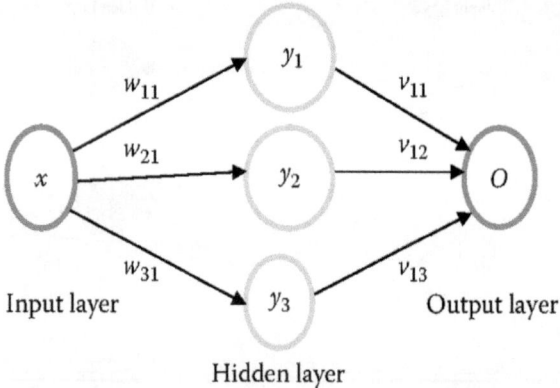

Fig. 1.3 A multilayer neural network architecture

1. Initialize the weights W from the input layer to the hidden layer and weights V from the hidden layer to the output layer. Then, we choose a learning parameter $\eta \in [0, 1]$ and error E_{max}. Initially, error is taken as $E = 0$.
2. Training steps start here. Outputs of the hidden layer and the output layer are computed as follows

$$f(w_1 x) \rightarrow y_1,$$
$$f(w_2 x) \rightarrow y_2,$$
$$f(w_3 x) \rightarrow y_3$$

and

$$f(v_1 y) \rightarrow o_1,$$

respectively. Here, w_j is the jth row of W, $j \in \{1, 2, 3\}$, v_1 is the first row of V and f is the activation function.
3. Error value is computed as follows

$$E = \frac{1}{2}(d_1 - o_1)^2,$$

where d_1 is the desired output and o_1 is the output of ANN.
4. The error signal term of the output layer is given by

$$\delta_{01} = (d_1 - o_1) f^{\Delta}(v_1 y).$$

The error signal term of the hidden layer is given by

$$\delta_{yj} = (1 - y_j) f(w_j x) \delta_{01} v_{1j},$$

where $o_1 = f(v_1 y)$.

5. Compute component of the error gradient vectors as follows

$$\frac{\partial E}{\Delta w_{j1}} = \delta_{yj}x_1, \quad j \in \{1, 2, 3\},$$

$$\frac{\partial E}{\Delta v_{1j}} = \delta_{01}y_j, \quad j \in \{1, 2, 3\}.$$

6. Weights are modified using the gradient descent method from the input layer to the hidden layer and from the hidden layer to the output layer as follows

$$w_{ji}^{n+1} = w_{ji}^n + \left(-\eta \frac{\partial E}{\Delta w_{ji}^n}\right),$$

$$v_{kj}^{n+1} = v_{kj}^n + \left(-\eta \frac{\partial E}{\Delta v_{kj}^n}\right),$$

where η is the learning parameter, n is the iteration step, E is the error function.
7. If $E = E_{max}$, then the session terminates. Otherwise, go to step 2 with $0 \rightarrow E$ and indicate new training.

1.3 Activation Functions

Suppose that $\mathbb{T}$ is a time scale with forward jump operator and delta differentiation operator σ and Δ, respectively.

Definition 1.1 An activation function or a transfer function is a function that acts upon the net (input) to get the output of the network.

The activation function transfers input signals into output signals. There are five types of activation functions.

1. Unit step function.
2. Piecewise linear function.
3. Gaussian function.
4. Sigmoid function.

Definition 1.2 The sigmoid function is defined as strictly increasing and continuously-differentiable function.

There are two types of sigmoid functions.

Definition 1.3 The unipolar sigmoid function is defined as follows

$$f(x, x_0) = \frac{1}{1 + e_{\ominus\lambda}(x, x_0)}, \quad x, x_0 \in \mathbb{T},$$

where $\lambda > 0$ is the slope of the function.

The output of the unipolar sigmoid function lies in the interval $[0, 1]$.

Definition 1.4 Bipolar sigmoid function is given by

$$f(x, x_0) = \frac{2}{1 + e_{\ominus\lambda}(x, x_0)} - 1, \quad x, x_0 \in \mathbb{T},$$

where $\lambda > 0$.

We have

$$f(x, x_0) = \frac{2 - 1 - e_{\ominus\lambda}(x, x_0)}{1 + e_{\ominus\lambda(x, x_0)}}$$

$$= \frac{1 - e_{\ominus\lambda}(x, x_0)}{1 + e_{\ominus\lambda}(x, x_0)}, \quad x, x_0 \in \mathbb{T}.$$

The output of bipolar sigmoid function lies in the interval $[-1, 1]$.

5. Tangent hyperbolic function.

Definition 1.5 The tangent hyperbolic function is given by

$$\tanh_f(x, x_0) = \frac{\sinh_f(x, x_0)}{\cosh_f(x, x_0)}$$

$$= \frac{e_f(x, x_0) - e_{-f}(x, x_0)}{e_f(x, x_0) + e_{-f}(x, x_0)}, \quad x, x_0 \in \mathbb{T},$$

where f is such that $\pm f \in \mathscr{R}$. Here with $\mathscr{R}$ we denote the regressive group on $\mathbb{T}$.

1.4 Examples

In this section, we will give some examples for the network training.

Example 1.1 Let $\mathbb{T} = \mathbb{R}$ and the activation function be the sigmoid activation function

$$f(t) = \frac{1}{1 + e^{-t}}, \quad t \in \mathbb{R}.$$

Consider a simple neural model without hidden layers. The training data are given by

$$t = (t_1, t_2)$$
$$= (0.1, 0.2),$$

the desired output is

$$y = 0.2,$$

the initial weights are

$$w = (w_1, w_2)$$
$$= (0.4, 0.1)$$

and the bias is

$$u = 1.78.$$

Since the considered model is without hidden layers, we have that

$$v = 1.$$

Firstly, we compute

$$z = t_1 w_1 + t_2 w_2 + u$$
$$= (0.1)(0.4) + (0.2)(0.1) + 1.78$$
$$= 0.04 + 0.02 + 1.78$$
$$= 0.06 + 1.78$$
$$= 1.84.$$

Then

$$f(z) = \frac{1}{1 + e^{-z}}$$
$$= \frac{1}{1 + e^{-1.84}}$$
$$= 0.863.$$

Thus,

$$y_a^1 = 0.863.$$

We see that the predicted output does not correspond to the desired output. Therefore we have to train the network to reduce the prediction error. For the prediction error, we have

$$E^1 = \frac{1}{2}(0.2 - 0.863)^2$$

$$= 0.22.$$

We see that the prediction error is very large. For this aim, we have to adjust the weights. This can be achieved using the formula

$$w_n^1 = w_o^1 + \eta(y - y_1)t, \tag{1.1}$$

where w_n and w_o represent new and old weights, respectively. We update the weights using the following

$$w_o^1 = \text{current weight}$$
$$= (1.78, 0.4, 0.1),$$
$$\eta = \text{network rate}$$
$$= 0.1,$$
$$y = \text{desitred output}$$
$$= 0.2,$$
$$t = \text{current input vector}$$
$$= (1, 0.1, 0.2).$$

Then, by the formula (1.1), we get

$$\begin{aligned}
w_n^1 &= (1.78, 0.4, 0.1) + 0.1(0.2 - 0.863)(1, 0.1, 0.2) \\
&= (1.78, 0.4, 0.1) + 0.1(-0.663)(1, 0.1, 0.2) \\
&= (1.78, 0.4, 0.1) + (-0.0663)(1, 0.1, 0.2) \\
&= (1.78, 0.4, 0.1) + (-0.0663, -0.00663, -0.01326) \\
&= (1.7137, 0.39337, 0.08674) \\
&\approx (1.714, 0.393, 0.087).
\end{aligned}$$

Then

$$\begin{aligned}
z^1 &= 0.1(0.393) + 0.02(0.087) + 1.714 \\
&= 0.0393 + 0.00174 + 1.714 \\
&= 1.75504
\end{aligned}$$

and

$$\begin{aligned}
f(z^1) &= \frac{1}{1 + e^{-1.75504}} \\
&= 0.85259
\end{aligned}$$

and

$$y_a^2 = 0.85259.$$

We see that

$$y_a^2 < y_a^1.$$

The prediction error is

$$E^2 = \frac{1}{2}(0.2 - 0.85259)^2$$
$$= 0.212$$
$$< E^1.$$

Again, we see that the prediction error is very large. We have to minimize it. We update the weights in the following manner.

$$w_o^2 = \text{current weight}$$
$$= w_n^1$$
$$= (1.714, 0.393, 0.087),$$
$$\eta = \text{network rate}$$
$$= 0.1,$$
$$y = \text{desired ouput}$$
$$= 0.2,$$
$$t = \text{current input vector}$$
$$= (1, 0.1, 0.2).$$

Then

$$w_n^2 = w_o^2 + \eta(y - y_a^2)t$$
$$= (1.714, 0.393, 0.087) + 0.1(0.2 - 0.85259)(1, 0.1, 0.2)$$
$$= (1.714, 0.393, 0.087) + 0.1(-0.65259)(1, 0.1, 0.2)$$
$$= (1.74, 0.393, 0.087) + (-0.065259)(1, 0.1, 0.2)$$
$$= (1.74, 0.393, 0.087) + (-0.065259, -0.0065259, -0.01305)$$
$$= (1.67474, 0.38647, 0.07395).$$

Hence,

$$z^2 = (0.38647)(0.1) + (0.07395)(0.2) + 1.67474$$
$$= 0.038647 + 0.01478 + 1.67474$$
$$= 1.72817$$

and

$$f(z^2) = \frac{1}{1 + e^{-1.72817}}$$
$$= 0.84918,$$

and

$$y_a^3 = 0.84918$$
$$< y_a^2.$$

The prediction error is

$$E^3 = \frac{1}{2}(0.2 - 0.84918)^2$$
$$= 0.2107$$
$$< E^2.$$

We continue in this way while we minimize the prediction error.

Example 1.2 Let $\mathbb{T} = \mathbb{Z}$. Then

$$\sigma(t) = t + 1,$$
$$\mu(t) = 1, \quad t \in \mathbb{T},$$

and

$$(\ominus 1)(t) = -\frac{1}{1 + \mu(t)}$$
$$= -\frac{1}{1 + 1}$$
$$= -\frac{1}{2},$$

and

$$e_{\ominus 1}(t, 0) = \prod_{\tau=0}^{t-1} \left(1 - \frac{1}{2}\right)$$
$$= \prod_{\tau=0}^{t-1} \frac{1}{2}$$
$$= \left(\frac{1}{2}\right)^t, \quad t \in \mathbb{T}.$$

Then, the sigmoid function is given by

$$f(t) = \frac{1}{1 + \left(\frac{1}{2}\right)^t}, \quad t \in \mathbb{T}.$$

Consider a simple neural network with one hidden layer. The training data is

$$t = 1,$$

the desired result is

$$y = 2,$$

the input weight is

$$w = 3,$$

and the hidden weight is

$$v = 2,$$

and the bias is

$$u = 2.$$

Firstly, we compute

$$
\begin{aligned}
z &= tw + u \\
&= 1 \cdot 3 + 2 \\
&= 5
\end{aligned}
$$

and

$$
\begin{aligned}
f(z) &= \frac{1}{1 + \left(\frac{1}{2}\right)^5} \\
&= 0.9697
\end{aligned}
$$

and

$$
\begin{aligned}
y_a^1 &= vf(z) \\
&= 2(0.9697) \\
&= 1.9394.
\end{aligned}
$$

We see that the predicted output does not correspond to the desired result. The prediction error is

$$
\begin{aligned}
E^1 &= \frac{1}{2}(2 - 1.9394)^2 \\
&= 0.00184.
\end{aligned}
$$

Therefore we have to minimize the error. For this aim, we will update the input weight, hidden weight and the bias. We have the following

$$w_n^1 = w_o^1 + \eta(y - y_a^1)t,$$
$$v_n^1 = v_o^1 + \eta(y - y_a^1)t,$$
$$u_n^1 = u_o^1 + \eta(y - y_a^1)t,$$

where

$$w_o^1 = 3,$$
$$v_o^1 = 2,$$
$$u_o^1 = 2,$$
$$t = 1,$$
$$\eta = 0.01.$$

Then

$$w_n^1 = 3 + 0.01(2 - 1.9394)$$
$$= 3.00061,$$

and

$$v_n^1 = 2 + 0.01(2 - 1.9394)$$
$$= 2.00061,$$

and

$$u_n^1 = 2 + 0.01(2 - 1.9394)$$
$$= 2.00061.$$

Hence,

$$z^1 = tw_n^1 + u_n^1$$
$$= 3.00061 + 2.00061$$
$$= 5.00122$$

and

$$f(z^1) = \frac{1}{1 + \left(\frac{1}{2}\right)^{5.00122}}$$
$$= 0.96972$$

and

$$y_a^2 = v_n^1 f(z^1)$$
$$= 2.00061(0.96972)$$
$$= 1.94003,$$

and the prediction error is

$$E^2 = \frac{1}{2}(2 - 1.94003)^2$$
$$= 0.00179$$
$$< E^1.$$

Now, we will update the input weight, hidden weight and bias in the following manner

$$w_n^2 = w_n^1 + \eta(y - y_a^2)t$$
$$= 3.00061 + 0.01(2 - 1.9394)$$
$$= 3.00122$$

and

$$v_n^2 = v_n^1 + \eta(y - y_a^2)t$$
$$= 2.00061 + 0.01(2 - 1.9394)$$
$$= 2.00122,$$

and

$$u_n^2 = u_n^1 + \eta(y - y_a^2)t$$
$$= 2.00061 + 0.01(2 - 1.9394)$$
$$= 2.00122.$$

Hence,

$$z^2 = tw_n^2 + u_n^2$$
$$= 3.00122 + 2.00122$$
$$= 5.00244$$

and

$$f(z^2) = \frac{1}{1 + \left(\frac{1}{2}\right)^{5.00244}}$$
$$= 0.96975,$$

and

$$y_a^3 = v_n^2 f(z^2)$$
$$= 2.00122(0.96975)$$
$$= 1.94068$$
$$> y_a^2,$$

and the prediction error is

$$E^3 = \frac{1}{2}(2 - 1.94068)^2$$
$$= 0.00176$$
$$< E^2.$$

We see that the prediction error decreases. We continue this process while we minimize the error.

Example 1.3 Let $\mathbb{T} = 2^{\mathbb{N}_0}$. Then

$$\sigma(t) = 2t,$$
$$\mu(t) = t, \quad t \in \mathbb{T}.$$

For $z \in \mathbb{R}$ such that

$$1 + tz \neq 0, \quad t \in \mathbb{T},$$

one has

$$(\ominus z)(t) = -\frac{z}{1 + \mu(t)z}$$
$$= -\frac{z}{1 + tz}, \quad t \in \mathbb{T},$$

and

$$e_{\ominus z}(t, 1) = e^{\int_1^t \frac{1}{\mu(\tau)} \log(1 + \mu(\tau)(\ominus z)(\tau)) \Delta \tau}$$
$$= e^{\int_1^t \frac{1}{t} \log(1 - \frac{\tau z}{1 + \tau z}) \Delta \tau}$$

$$= e^{\int_1^t \frac{1}{\tau} \log \frac{1}{1+\tau z} \Delta \tau}$$

$$= e^{\sum_{\tau=1}^{\tau=\frac{t}{2}} \log \frac{1}{1+\tau z}}$$

$$= \prod_{\tau=1}^{\frac{t}{2}} \frac{1}{1+\tau z}, \quad t \in \mathbb{T},$$

and the sigmoid function is

$$f(z) = \frac{1}{1 + \prod_{\tau=1}^{\frac{t}{2}} \frac{1}{1+\tau z}}, \quad t \in \mathbb{T}.$$

For $t = 2$, we get

$$f(z) = \frac{1}{1 + \frac{1}{1+z}}.$$

Consider a simple neural network with a hidden layer. The training data, desired output, input weight, hidden weight, bias and network rate are as follows

$$t = 2,$$
$$y = 2,$$
$$w = 2,$$
$$v = 2,$$
$$u = 2,$$
$$\eta = 0.01.$$

Firstly, compute

$$z = tw + u$$
$$= 2 \cdot 2 + 2$$
$$= 4 + 2$$
$$= 6$$

and

$$f(z) = \frac{1}{1 + \frac{1}{1+6}}$$
$$= 0.875,$$

and

$$y_a^1 = vf(z)$$
$$= 2(0.875)$$
$$= 1.750.$$

We see that the predicted output does not coincide with the desired result. The predicted error is

$$E^1 = \frac{1}{2}(y - y_a^1)^2$$
$$= \frac{1}{2}(2 - 1.75)^2$$
$$= 0.03125.$$

We will minimize the predicted error. For this aim we will update the network data in the following way

$$w_n^1 = w_o^1 + \eta(y - y_a^1)t,$$
$$v_n^1 = v_o^1 + \eta(y - y_a^1)t,$$
$$u_n^1 = u_o^1 + \eta(y - y_a^1)t,$$

where

$$w_o^1 = 2,$$
$$v_o^1 = 2,$$
$$u_o^1 = 2.$$

We have

$$w_n^1 = 2 + 0.01(2 - 1.75)2$$
$$= 2 + 0.02(0.25)$$
$$\approx 2.0015$$

and

$$v_n^1 = 2 + 0.01(2 - 1.75)2$$
$$= 2 + 0.02(0.25)$$
$$\approx 2.0015,$$

and

$$u_n^1 = 2 + 0.01(2 - 1.75)2$$
$$= 2 + 0.02(0.25)$$

$$\approx 2.0015.$$

Then

$$z^1 = tw_n^1 + u_n^1$$
$$= 2 \cdot 2.0015 + 2.0015$$
$$= 4.0030 + 2.0015$$
$$= 6.0045$$

and

$$f(z^1) = \frac{1}{1 + \frac{1}{1+6.0045}}$$
$$= 0.87507.$$

Hence,

$$y_a^2 = v_n^1 f(z^1)$$
$$= 2.0015 \cdot 0.87507$$
$$= 1.75145$$
$$> y_a^1$$

and the prediction error is

$$E^2 = \frac{1}{2}(y - y_a^1)^2$$
$$= \frac{1}{2}(2 - 1.75145)^2$$
$$= 0.03089$$
$$< E^1.$$

Again the predicted output does not correspond to the desired result. Therefore we update the network data as follows

$$w_n^2 = w_n^1 + \eta(y - y_a^2)t$$
$$= 2.0015 + 0.01(2 - 1.75145)2$$
$$= 2.0015 + 0.02(2 - 1.75145)$$
$$= 2.00647$$

and

$$v_n^2 = v_n^1 + \eta(y - y_a^2)t$$
$$= 2.0015 + 0.01(2 - 1.75145)2$$
$$= 2.0015 + 0.02(2 - 1.75145)$$
$$= 2.00647,$$

and

$$u_n^2 = u_n^1 + \eta(y - y_a^2)t$$
$$= 2.0015 + 0.01(2 - 1.75145)2$$
$$= 2.0015 + 0.02(2 - 1.75145)$$
$$= 2.00647.$$

Hence,

$$z^2 = tw_n^2 + u_n^2$$
$$= 2(2.00647) + 2.00647$$
$$= 3(2.00647)$$
$$= 6.01941$$

and

$$f(z^2) = \frac{1}{1 + \frac{1}{1+6.01941}}$$
$$= 0.8753$$

and

$$y_a^3 = v^2 f(z^2)$$
$$= 2.00647(0.8753)$$
$$= 1.75626$$
$$> y_a^2.$$

The predicted error is

$$E^3 = \frac{1}{2}(y - y_a^3)^2$$
$$= \frac{1}{2}(2 - 1.75626)^2$$
$$= 0.0297$$

$$< E^2.$$

We see that the prediction error decreases. We continue this process while we minimize the predicted error.

Chapter 2
Multilayer Artificial Neural Networks

In this chapter, the multilayer artificial neural network (ANN) model has been introduced for solving of dynamic equations on arbitrary time scales. A multilayer ANN model with one input layer containing a single node, a hidden layer with m nodes, and one output node are considered. The ANN trial solution of the considered dynamic equations is represented as a sum of two terms. The first term satisfies the initial or boundary conditions. The second term contains the output of the ANN with adjustable parameters. The feed-forward neural network model and unsupervised error back-propagation algorithm have been investigated. Modification of network parameters has been done without the use of any optimization technique.

Suppose that $\mathbb{T}$ is a time scale with forward jump operator and delta differentiation operator σ and Δ, respectively.

2.1 Learning Algorithm for Multilayer ANN Model

First, we consider a general form of dynamic equations on $\mathbb{T}$

$$F\left(t, y(t), y^{\Delta}(t), \ldots, y^{\Delta^n}(t)\right) = 0, \quad t \in I \subset \mathbb{T}, \qquad (2.1)$$

subject to certain initial or boundary conditions. Here F is a given function and y is the unknown.

Let $y_a(t, p)$ denote the ANN trial solution of (2.1) with adjustable parameters $p = (p_1, \ldots, p_n)$ (weights or biases). Then, the Eq. (2.1) can be rewritten in the form

$$F\left(t, y_a(t, p), y_a^{\Delta}(t, p), \ldots, y_a^{\Delta^n}(t, p)\right) = 0, \quad t \in I.$$

The trial solution $y_a(t, p)$ of feed-forward neural network with input t and parameters p may be written in the form

© The Author(s), under exclusive license to Springer Nature Switzerland AG 2025
S. Georgiev, *Neural Network Methods for Dynamic Equations on Time Scales*,
SpringerBriefs in Computational Intelligence,
https://doi.org/10.1007/978-3-031-85056-1_2

$$y_a(t, p) = A(t) + G(t, N(t, p)),$$

where $A(t)$ satisfies the initial or boundary conditions and contains no adjustable parameters, whereas $N(t, p)$ is the output of feed-forward neural network with parameters p and input t. The term $G(t, N(t, p))$ makes no contributions in the initial or boundary conditions but it is the output of the neural network model whose weights and biases are adjusted to minimize the error function to obtain the trial ANN solution $y_a(t, p)$.

Obviously, the network output $N(t, p)$ is given by

$$N(t, p) = \sum_{j=1}^{m} v_j s(w_j t + u_j),$$

where w_j denotes the weight from the input unit to the hidden unit j, v_j denotes the weight from the hidden unit j to the output unit, u_j is a biases, and $s(w_j t + u_j)$ is the activation function, $j \in \{1, \ldots, m\}$.

In the training model, we start with the given weights and biases and train the model to modify the weights in the given interval of the problem. Our goal is to minimize the error function.

2.2   Formulation for Initial Value Problems

In this section, we will describe the learning algorithm for first order, second order and nth order initial value problems (IVP).

2.2.1   First Order Dynamic Equations

Consider the first order IVP

$$\begin{aligned} y^\Delta &= f(t, y), \quad t \in [t_0, T], \\ y(t_0) &= y_0, \end{aligned} \tag{2.2}$$

where $t_0, T \in \mathbb{T}$, $t_0 < T$, $y_0 \in \mathbb{R}$, f is a given function. In this case, the ANN trial solution is given by

$$y_a(t, p) = y_0 + h_1(t, t_0)N(t, p), \quad t \in [t_0, T], \tag{2.3}$$

where $N(t, p)$ is the output of feed-forward network with input data t and parameters p. We differentiate (2.3) and we get

$$y_a^\Delta(t, p) = N(t, p) + h_1(\sigma(t), t_0)N^\Delta(t, p)$$
$$= N(\sigma(t), p) + h_1(t, t_0)N^\Delta(t, p), \quad t \in [t_0, T].$$

The error function in this case is given by

$$E(t, p) = \sum_{j=1}^{h} \frac{1}{2} \left(y_a^\Delta(t_j, p) - f(t_j, y_a(t_j, p)) \right)^2, \quad t \in [t_0, T].$$

2.2.2 Second Order Dynamic Equations

Here, we suppose that the graininess function is delta differentiable on $[t_0, T]$. Consider the IVP

$$y^{\Delta^2} = f\left(t, y, y^\Delta\right), \quad t \in [t_0, T],$$
$$y(t_0) = y_0, \quad\quad\quad\quad\quad\quad (2.4)$$
$$y^\Delta(t_0) = y_1,$$

where $y_0, y_1 \in \mathbb{R}$ and f is a given function. The ANN trial solution of the IVP (2.4) is given by

$$y_a(t, p) = y_0 + h_1(t, t_0)y_1 + h_2(t, t_0)N(t, p), \quad t \in [t_0, T],$$

where $N(t, p)$ is the output of the ANN with input t and parameters p. We differentiate $y_a(t, p)$ with respect to t and we find

$$y_a^\Delta(t, p) = y_1 + h_1(t, t_0)N(t, p) + h_2(\sigma(t), t_0)N^\Delta(t, p)$$
$$= y_1 + h_1(t, t_0)N(t, p) + (h_2(t, t_0)$$
$$+ \mu(t)h_1(t, t_0))N^\Delta(t, t_0), \quad t \in [t_0, T],$$

and

$$y_a^{\Delta^2}(t, p) = N(t, p) + h_1(\sigma(t), t_0)N^\Delta(t, p)$$
$$+ h_1(t, t_0)N^\Delta(t, p) + h_2(\sigma(t), t_0)N^{\Delta^2}(t, p)$$
$$+ \mu^\Delta(t)h_1(t, t_0)N^\Delta(t, p)$$
$$+ \mu(\sigma(t))N^\Delta(t, p) + \mu(\sigma(t))h_1(\sigma(t), t_0)N^{\Delta^2}(t, p)$$
$$= N(t, p)$$
$$+ \left(h_1(\sigma(t), t_0) + h_1(t, t_0) + \mu^\Delta(t)h_1(t, t_0) + \mu(\sigma(t)) \right) N^\Delta(t, p)$$
$$+ (h_2(\sigma(t), t_0) + \mu(\sigma(t))h_1(\sigma(t), t_0)) N^{\Delta^2}(t, p), \quad t \in [t_0, T].$$

The error function in this case is as follows

$$E(t, p) = \sum_{j=1}^{h} \frac{1}{2} \left(y_a^{\Delta}(t_j, p) - f\left(t_j, y_a(t_j, p), y_a^{\Delta}(t_j, p)\right) \right)^2, \quad t \in [t_0, T].$$

2.2.3 nth Order Dynamic Equations

Here, we suppose that σ is n times delta differentiable on $[t_0, T]$. Consider the following IVP

$$\begin{aligned}
y^{\Delta^n} &= f\left(t, y, y^{\Delta}, \dots, y^{\Delta^{n-1}}\right), \quad t \in [t_0, T], \\
y(t_0) &= y_0, \\
y^{\Delta}(t_0) &= y_1, \\
&\vdots \\
y^{\Delta^{n-1}}(t_0) &= y_{n-1},
\end{aligned}$$
(2.5)

where $y_j \in \mathbb{R}$, $j \in \{0, 1, \dots, n-1\}$, and f is a given function. The ANN trial solution for the considered problem is given by

$$y_a(t, p) = \sum_{j=0}^{n-1} y_j h_j(t, t_0) + h_n(t, t_0) N(t, p), \quad t \in [t_0, T],$$
(2.6)

where $N(t, p)$ is the output of the ANN with input t and parameters p.

Let $S_k^{(n)}$ denote the set of all strings of length n containing σ exactly k times and Δ exactly $n - k$ times. Suppose

$$f^{\Lambda} \quad \text{exists for any} \quad \Lambda \in S_k^{(l)}, \quad l \in \{0, 1, \dots, n\}.$$

Then, applying the Leibniz formula, for the kth delta derivative of (2.6) we have the following representation

$$y_a^{\Delta^k}(t, p) = \sum_{j=k}^{n-1} y_j h_{j-k}(t, t_0) + \sum_{l=0}^{k} \left(\sum_{\Lambda \in S_l^{(k)}} h_n^{\Lambda}(t, t_0) \right) N^{\Delta^k}(t, p),$$

$t \in [t_0, T]$. The error function is given by

$$E(t, p) = \sum_{j=1}^{h} \frac{1}{2} \left(y_a^{\Delta^n}(t_j, p) - f\left(t_j, y_a(t_j, p), y_a^{\Delta}(t_j, p), \dots, y_a^{\Delta^{n-1}}(t_j, p)\right) \right)^2,$$

$$t \in [t_0, T].$$

2.3 Formulation for Boundary Value Problems

In this section, we will formulate the learning algorithm for some classes boundary value problems (BVP) for first order, second order and nth order dynamic equations. Suppose that $t_0, T \in \mathbb{T}, t_0 < T$.

2.3.1 First Order Dynamic Equations

Consider the following boundary value problem

$$
\begin{aligned}
y^\Delta &= f(t, y), \quad t \in [t_0, T], \\
y(t_0) &= C, \\
y(T) &= D,
\end{aligned}
\tag{2.7}
$$

where $C, D \in \mathbb{R}$ are given constants and f is a given function. Firstly, we will search a function

$$
A(t) = c_1 h_1(t, t_0) + c_2, \quad t \in [t_0, T],
$$

where c_1 and c_2 are constants which will be determined by the boundary conditions

$$
\begin{aligned}
A(t_0) &= C, \\
A(T) &= D.
\end{aligned}
$$

We have the following

$$
\begin{aligned}
C &= A(t_0) \\
&= c_2
\end{aligned}
$$

and

$$
\begin{aligned}
D &= A(T) \\
&= c_1 h_1(T, t_0) + C,
\end{aligned}
$$

whereupon

$$
c_1 h_1(T, t_0) = D - C,
$$

or

$$
c_1 = \frac{D - C}{h_1(T, t_0)}.
$$

Therefore

$$A(t) = \frac{D - C}{h_1(T, t_0)} h_1(t, t_0) + C, \quad t \in [t_0, T].$$

Then the ANN trial solution of the considered BVP is given by

$$y_a(t, p) = \frac{D - C}{h_1(T, t_0)} h_1(t, t_0) + C + h_2(t, t_0) h_2(t, T) N(t, p), \quad t \in [t_0, T],$$

where $N(t, p)$ is the output of the ANN with input t and parameters p. For the first derivative of $y_a(t, p)$ we have the following representation

$$y_a^\Delta(t, p) = \frac{D - C}{h_1(T, t_0)} + h_1(t, t_0) h_2(t, T) N(t, p) + h_2(\sigma(t), t_0) h_1(t, T) N(t, p)$$
$$+ h_2(\sigma(t), t_0) h_2(\sigma(t), T) N^\Delta(t, p), \quad t \in [t_0, T].$$

The error function is given by

$$E(t, p) = \sum_{j=1}^{h} \frac{1}{2} \left(y_a^\Delta(t_j, p) - f(t_j, y(t_j, p)) \right)^2, \quad t \in [t_0, T].$$

Now, consider the following BVP

$$y^\Delta = f(t, y), \quad t \in [t_0, T],$$
$$My(t_0) + Ry(\sigma(T)) = 0, \tag{2.8}$$

where $M, R \in \mathbb{R}$ are given constants and $f : [t_0, T] \times \mathbb{R} \to \mathbb{R}$ is a given function. Firstly, we will find a function $A : [t_0, T] \to \mathbb{R}$ in the form

$$A(t) = c_1 + c_2 h_1(t, t_0), \quad t \in [t_0, T],$$

where $c_1, c_2 \in \mathbb{R}$ will be determined so that

$$MA(t_0) + RA(\sigma(T)) = 0.$$

We have

$$A(t_0) = c_1,$$
$$A(\sigma(T)) = c_1 + c_2 h_1(\sigma(T), t_0).$$

Hence,

$$0 = MA(t_0) + RA(\sigma(T))$$
$$= Mc_1 + Rc_1 + Rc_2 h_1(\sigma(T), t_0)$$
$$= (M + R)c_1 + Rc_2 h_1(\sigma(T), t_0),$$

or

$$Rh_1(\sigma(T), t_0)c_2 = -(M + R)c_1,$$

and

$$c_2 = -\frac{(M + R)c_1}{Rh_1(\sigma(T), t_0)}.$$

Then

$$A(t) = c_1 - \frac{(M + R)c_1}{Rh_1(\sigma(T), t_0)}h_1(t, t_0), \quad t \in [t_0, T].$$

Therefore an ANN trial solution for the considered BVP is given by the expression

$$\begin{aligned}
y_a(t, p) &= A(t) + h_2(t, t_0)h_2(t, \sigma(T))N(t, p) \\
&= c_1 - \frac{(M + R)c_1}{Rh_1(\sigma(T), t_0)}h_1(t, t_0) \\
&\quad + h_2(t, t_0)h_2(t, \sigma(T))N(t, p), \quad t \in [t_0, T],
\end{aligned}$$

where $N(t, p)$ is the output of the ANN with input t and parameters p and c_1 is an arbitrary real constant. For its first derivative we have the following

$$\begin{aligned}
y_a^\Delta(t, p) = &-\frac{(M + R)c_1}{Rh_1(\sigma(T), t_0)} + h_1(t, t_0)h_2(t, \sigma(T))N(t, p) \\
&+ h_2(\sigma(t), t_0)h_1(t, \sigma(T))N(t, p) + h_2(\sigma(t), t_0)h_2(\sigma(t), \sigma(T))N^\Delta(t, p), \\
&\qquad\qquad\qquad\qquad\qquad\qquad\qquad\qquad\qquad t \in [t_0, T].
\end{aligned}$$

The error function is given by

$$E(t, p) = \sum_{j=1}^{h} \frac{1}{2}\left(y_a^\Delta(t_j, p) - f(t_j, y(t_j, p))\right)^2, \quad t \in [t_0, T].$$

Exercise 2.1 Find an ANN trial solution for the following BVP

$$y^\Delta = f(t, y), \quad t \in [t_0, T],$$
$$y(t_0) = y(\sigma(T)),$$

where $f : [t_0, T] \times \mathbb{R} \to \mathbb{R}$ is a given function.
Answer 2.1

$$y_a(t, p) = c + h_2(t, t_0)h_2(t, \sigma(T))N(t, p), \quad t \in [t_0, T],$$

where c is an arbitrary real constant, $N(t, p)$ is the output of the ANN with input t and parameters p.

2.3.2   Second Order Dynamic Equations

In this section, assume that the forward jump operator σ is differentiable. Consider the following BVP

$$
\begin{aligned}
y^{\Delta^2} &= f\left(t, y, y^{\Delta}\right), \quad t \in [t_0, T], \\
y(t_0) &= C, \\
y^{\Delta}(T) &= D,
\end{aligned}
\tag{2.9}
$$

where C and D are given real constants, $f : [t_0, T] \times \mathbb{R} \times \mathbb{R} \to \mathbb{R}$ is a given function. We will find a function $A : [t_0, T] \to \mathbb{R}$ in the form

$$
A(t) = c_1 + c_2 h_1(t, t_0), \quad t \in [t_0, T],
$$

where c_1 and c_2 are real constants which will be determined by the conditions

$$
\begin{aligned}
A(t_0) &= C, \\
A^{\Delta}(T) &= D.
\end{aligned}
$$

We have

$$
\begin{aligned}
A(t_0) &= c_1 \\
&= C
\end{aligned}
$$

and

$$
A^{\Delta}(t) = c_2, \quad t \in [t_0, T],
$$

and then

$$
D = C_2.
$$

Therefore

$$
A(t) = C + D h_1(t, t_0), \quad t \in [t_0, T].
$$

Hence, an ANN trial solution for the considered BVP is given by the expression

$$
y_a(t, p) = C + D h_1(t, t_0) + h_2(t, t_0) h_2(t, T) N(t, p), \quad t \in [t_0, T],
$$

where $N(t, p)$ is the output of the ANN with input t and parameters p. For its first and second derivatives we have the following representations

$$
\begin{aligned}
y_a^{\Delta}(t, p) = {} & D + h_1(t, t_0) h_2(t, T) N(t, p) \\
& + h_2(\sigma(t), t_0) h_1(t, T) N(t, p) \\
& + h_2(\sigma(t), t_0) h_2(\sigma(t), T) N^{\Delta}(t, p), \quad t \in [t_0, T],
\end{aligned}
$$

and

$$
\begin{aligned}
y_a^{\Delta^2}(t, p) = {} & h_2(t, T)N(t, p) + h_1(\sigma(t), t_0)h_1(t, T)N(t, p) \\
& + h_1(\sigma(t), t_0)h_2(\sigma(t), T)N^{\Delta}(t, p) \\
& + (h_2(\sigma(\cdot), t_0))^{\Delta}(t)h_1(t, T)N(t, p) \\
& + h_2\left(\sigma^2(t), t_0\right)N(t, p) + h_2\left(\sigma^2(t), t_0\right)h_1(\sigma(t), T)N^{\Delta}(t, p) \\
& + (h_2(\sigma(\cdot), t_0))^{\Delta}(t)h_2(\sigma(t), T)N^{\Delta}(t, p) \\
& + h_2\left(\sigma^2(t), t_0\right)(h_2(\sigma(\cdot), t_0))^{\Delta}(t)N^{\Delta}(t, p) \\
& + h_2\left(\sigma^2(t), t_0\right)h_2\left(\sigma^2(t), T\right)N^{\Delta^2}(t, p), \quad t \in [t_0, T].
\end{aligned}
$$

The error function is given by

$$
E(t, p) = \sum_{j=1}^{h} \frac{1}{2}\left(y_a^{\Delta^2}(t_j, p) - f\left(t_j, y(t_j, p), y_a^{\Delta}(t_j, p)\right)\right)^2, \quad t \in [t_0, T].
$$

Now, consider the following BVP

$$
\begin{aligned}
y^{\Delta^2} &= f(t, y, y^{\Delta}), \quad \in [t_0, T], \\
\alpha y(t_0) - \beta y^{\Delta}(t_0) &= 0, \\
\gamma y(\sigma^2(t_0)) + \delta y^{\Delta}(\sigma(T)) &= 0,
\end{aligned}
\tag{2.10}
$$

where α, β, γ and δ are real constants such that

$$
\alpha(\gamma h_1(\sigma^2(t_0), t_0) + \delta) + \gamma\beta \neq 0, \tag{2.11}
$$

$$
\alpha \neq 0, \tag{2.12}
$$

$$
\gamma h_2(\sigma^2(t_0), t_0) + \delta h_1(\sigma(T), t_0) \neq 0, \tag{2.13}
$$

and $f : [t_0, T] \times \mathbb{R} \times \mathbb{R} \to \mathbb{R}$ is a given function. We will find a function $A : [t_0, T] \to \mathbb{R}$ in the following form

$$
A(t) = c_1 + c_2 h_1(t, t_0), \quad t \in [t_0, T], \tag{2.14}
$$

where c_1 and c_2 are real constants which will be determined by the conditions

$$
\begin{aligned}
\alpha A(t_0) - \beta A^{\Delta}(t_0) &= 0, \\
\gamma A(\sigma^2(t_0)) + \delta A^{\Delta}(\sigma(T)) &= 0.
\end{aligned}
\tag{2.15}
$$

We have

$$
A^{\Delta}(t) = c_2, \quad t \in [t_0, T],
$$

and

$$A(t_0) = c_1,$$
$$A^{\Delta}(t_0) = c_2,$$
$$A(\sigma^2(t_0)) = c_1 + c_2 h_1(\sigma^2(t_0), t_0),$$
$$A^{\Delta}(\sigma(T)) = c_2.$$

Thus, we get the system

$$\alpha c_1 - \beta c_2 = 0$$
$$\gamma(c_1 + c_2 h_1(\sigma^2(t_0), t_0)) + \delta c_2 = 0,$$

or

$$\alpha c_1 - \beta c_2 = 0$$
$$\gamma c_1 + (\gamma h_1(\sigma^2(t_0), t_0) + \delta)c_2 = 0$$

Since (2.11) holds, the last system has only the zero solution and the function A in the form (2.14) is not appropriate for our aims. Now, we will search a function $A : [t_0, T] \to \mathbb{R}$ in the form

$$A(t) = c_1 + c_2 h_1(t, t_0) + c_3 h_2(t, t_0), \quad t \in [t_0, T],$$

where c_1, c_2 and c_3 are constants that will be determined by the conditions (2.15). We have

$$A^{\Delta}(t) = c_2 + c_3 h_1(t, t_0), \quad t \in [t_0, T],$$

and

$$A(t_0) = c_1,$$
$$A^{\Delta}(t_0) = c_2,$$
$$A(\sigma^2(t_0)) = c_1 + c_2 h_1(\sigma^2(t_0), t_0) + c_3 h_2(\sigma^2(t_0), t_0),$$
$$A^{\Delta}(\sigma(T)) = c_2 + c_3 h_1(\sigma(T), t_0).$$

From here, we get the following system

$$\alpha c_1 - \beta c_2 = 0$$
$$\gamma(c_1 + c_2 h_1(\sigma^2(t_0), t_0) + c_3 h_2(\sigma^2(t_0), t_0)) + \delta(c_2 + c_3 h_1(\sigma(T), t_0)) = 0,$$

or

$$\alpha c_1 - \beta c_2 = 0,$$

$$\gamma c_1 + c_2(\gamma h_1(\sigma^2(t_0), t_0) + \delta) + c_3(\gamma h_2(\sigma^2(t_0), t_0) + \delta h_1(\sigma(T), t_0)) = 0.$$

Now, using (2.12), we find

$$c_1 = \frac{\beta}{\alpha} c_2$$

and

$$\frac{\gamma \beta}{\alpha} c_2 + c_2(\gamma h_1(\sigma^2(t_0), t_0) + \delta) + c_3(\gamma h_2(\sigma^2(t_0), t_0) + \delta h_1(\sigma(T), t_0)) = 0,$$

or

$$-c_2(\gamma \beta + \alpha \gamma h_1(\sigma^2(t_0), t_0) + \alpha \delta) = c_3 \alpha(\gamma h_2(\sigma^2(t_0), t_0) + \delta h_1(\sigma(T), t_0)),$$

whereupon, applying (2.12) and (2.13), we obtain

$$c_3 = -c_2 \frac{\gamma \beta + \alpha \gamma h_1(\sigma^2(t_0), t_0) + \alpha \delta}{\alpha(\gamma h_2(\sigma^2(t_0), t_0) + \delta h_1(\sigma(T), t_0))},$$

where c_2 is an arbitrary real constant. An ANN trial solution of the considered BVP is given by

$$y_a(t, p) = \frac{\beta}{\alpha} c_2 + c_2 h_1(t, t_0) - c_2 \frac{\gamma \beta + \alpha \gamma h_1(\sigma^2(t_0), t_0) + \alpha \delta}{\alpha(\gamma h_2(\sigma^2(t_0), t_0) + \delta h_1(\sigma(T), t_0))} h_2(t, t_0)$$
$$+ h_2(t, t_0) h_1(t, \sigma^2(t_0)) h_2(t, \sigma(T)) N(t, p), \quad t \in [t_0, T],$$

where $N(t, p)$ is the output of the ANN with input t and parameters p. For its first and second derivatives we have the following representations

$$y_a^\Delta(t, p) = c_2 - c_2 \frac{\gamma \beta + \alpha \gamma h_1(\sigma^2(t_0), t_0) + \alpha \delta}{\alpha(\gamma h_2(\sigma^2(t_0), t_0) + \delta h_1(\sigma(T), t_0))} h_1(t, t_0)$$
$$+ h_1(t, t_0) h_1(t, \sigma^2(t_0)) h_2(t, \sigma(T)) N(t, p)$$
$$+ h_2(\sigma(t), t_0) h_2(t, \sigma(T)) N(t, p)$$
$$+ h_2(\sigma(t), t_0) h_1(\sigma(t), \sigma^2(t_0)) h_1(t, \sigma(T)) N(t, p)$$
$$+ h_2(\sigma(t), t_0) h_1(\sigma(t), \sigma^2(t_0)) h_2(\sigma(t), \sigma(T)) N^\Delta(t, p), \quad t \in [t_0, T],$$

and

$$
\begin{aligned}
y_a^{\Delta^2}(t, p) = &-c_2 \frac{\gamma\beta + \alpha\gamma h_1(\sigma^2(t_0), t_0) + \alpha\delta}{\alpha(\gamma h_2(\sigma^2(t_0), t_0) + \delta h_1(\sigma(T), t_0))} \\
&+ h_1(t, \sigma^2(t_0)) h_2(t, \sigma(T)) N(t, p) \\
&+ h_1(\sigma(t), t_0) h_2(t, \sigma(T)) N(t, p) \\
&+ h_1(\sigma(t), t_0) h_1(\sigma(t), \sigma^2(t_0)) h_1(t, \sigma(T)) N(t, p) \\
&+ h_1(\sigma(t), t_0) h_1(\sigma(t), \sigma^2(t_0)) h_2(\sigma(t), \sigma(T)) N^{\Delta}(t, p) \\
&+ (h_2(\sigma(\cdot), t_0))^{\Delta}(t) h_2(t, \sigma(T)) N(t, p) \\
&+ h_2(\sigma^2(t), t_0) h_1(t, \sigma(T)) N(t, p) \\
&+ h_2(\sigma^2(t), t_0) h_2(\sigma(t), \sigma(T)) N^{\Delta}(t, p) \\
&+ (h_2(\sigma(\cdot), t_0))^{\Delta}(t) h_1(\sigma(t), \sigma^2(t_0)) h_1(t, \sigma(T)) N(t, p) \\
&+ h_2(\sigma^2(t), t_0) (h_1(\sigma(\cdot), \sigma^2(t_0)))^{\Delta}(t) h_1(t, \sigma(T)) N(t, p) \\
&+ h_2(\sigma^2(t), t_0) h_1(\sigma^2(t), \sigma^2(t_0)) N(t, p) \\
&+ h_2(\sigma^2(t), t_0) h_1(\sigma^2(t), \sigma^2(t_0)) h_1(\sigma(t), \sigma(T)) N^{\Delta}(t, p) \\
&+ (h_2(\sigma(\cdot), t_0))^{\Delta}(t) h_1(\sigma(t), \sigma^2(t_0)) h_2(\sigma(t), \sigma(T)) N^{\Delta}(t, p) \\
&+ h_2(\sigma^2(t), t_0) (h_1(\sigma(\cdot), \sigma^2(t_0)))^{\Delta}(t) h_2(\sigma(t), \sigma(T)) N^{\Delta}(t, p) \\
&+ h_2(\sigma^2(t), t_0) h_1(\sigma^2(t), \sigma^2(t_0)) (h_2(\sigma(\cdot), \sigma(T)))^{\Delta}(t) N^{\Delta}(t, p) \\
&+ h_2(\sigma^2(t), t_0) h_1(\sigma^2(t), \sigma^2(t_0)) h_2(\sigma^2(t), \sigma(T)) N^{\Delta^2}(t, p), \quad t \in [t_0, T].
\end{aligned}
$$

The error function is given by

$$
E(t, p) = \sum_{j=1}^{h} \frac{1}{2} \left(y_a^{\Delta^2}(t_j, p) - f\left(t_j, y(t_j, p), y_a^{\Delta}(t_j, p)\right) \right)^2, \quad t \in [t_0, T].
$$

Example 2.1 Let $\mathbb{T} = 2\mathbb{Z}$. Consider the following BVP

$$
y^{\Delta^2} = \frac{1}{1 + t^2} \left(y^{\Delta}\right)^3, \quad t \in [0, 16],
$$
$$
y(0) - y^{\Delta}(0) = 0,
$$
$$
2y(4) + 3y^{\Delta}(18) = 0.
$$

Here

$$
\begin{aligned}
t_0 &= 0, \\
T &= 16, \\
\alpha &= 1, \\
\beta &= 1, \\
\gamma &= 2, \\
\delta &= 3
\end{aligned}
$$

and

$$h_1(t, s) = t - s,$$
$$h_2(t, s) = \frac{t^2}{2} - t - \frac{s^2}{2} + s - s(t - s), \quad t, s \in \mathbb{T}.$$

Then $\alpha \neq 0$ and

$$\alpha(\gamma h_1(\sigma^2(0), 0) + \delta) + \gamma\beta = 2 \cdot 4 + 3 + 2$$
$$= 13$$
$$\neq 0,$$

and

$$\gamma h_2(\sigma^2(0), 0) + \delta h_1(\sigma(16), 0) = 2h_2(4, 0) + 3h_1(18, 0)$$
$$= 2\left(\frac{4^2}{2} - 4\right) + 3(18 - 0)$$
$$= 2(8 - 4) + 54$$
$$= 8 + 54$$
$$= 62.$$

Thus, the conditions (2.11), (2.12) and (2.13) hold. Therefore an ANN trial solution of the considered BVP is given by

$$y_a(t, p) = c_2 + c_2 h_1(t, 0) - c_2 \frac{2 + 2h_1(4, 0) + 3}{2h_2(4, 0) + 3h_1(18, 0)} h_2(t, 0)$$
$$+ h_2(t, 0)h_1(t, 4)h_2(t, 18)N(t, p)$$
$$= c_2 + c_2 t - c_2 \frac{2 + 8 + 3}{62}\left(\frac{t^2}{2} - t\right)$$
$$+ \left(\frac{t^2}{2} - t\right)(t - 4)\left(\frac{t^2}{2} - t - \frac{324}{2} + 18 - 18(t - 18)\right)N(t, p)$$
$$= c_2 + c_2 t - c_2 \frac{13}{62}\left(\frac{t^2}{2} - t\right)$$
$$+ \left(\frac{t^2}{2} - t\right)(t - 4)\left(\frac{t^2}{2} - t + 162 + 18 - 18t\right)N(t, p)$$
$$= c_2 + c_2 t - c_2 \frac{13}{62}\left(\frac{t^2}{2} - t\right)$$
$$+ \left(\frac{t^2}{2} - t\right)(t - 4)\left(\frac{t^2}{2} - 19t + 180\right)N(t, p), \quad t \in [0, 16],$$

where c_2 is an arbitrary real constant and $N(t, p)$ is the output of the ANN with input t and parameters p.

Exercise 2.2 Find an ANN trial solution to the following BVP

$$y^{\Delta^2} = f(t, y, y^\Delta), \quad t \in [t_0, T],$$
$$y(t_0) = 0,$$
$$y^\Delta(\sigma(T)) = 0,$$

where $f : [t_0, T] \times \mathbb{R} \times \mathbb{R} \to \mathbb{R}$ is a given function.

Answer 2.2

$$y_a(t, p) = c_2 h_1(t, t_0) - c_2 \frac{1}{h_1(\sigma(T), t_0)} h_2(t, t_0)$$
$$+ h_2(t, t_0) h_1(t, \sigma^2(t_0)) h_2(t, \sigma(T)) N(t, p), \quad t \in [t_0, T],$$

where c_2 is an arbitrary real constant and $N(t, p)$ is the output of the ANN with input t and parameters p.

2.3.3 nth Order Dynamic Equations

Suppose that $n \geq 4$, $n \in \mathbb{N}$. In this section, consider the following nth order BVP

$$y^{\Delta^n} = f\left(t, y, y^\Delta, \ldots, y^{\Delta^{n-1}}\right), \quad t \in [t_0, T],$$
$$y^{\Delta^j}(t_0) = \lambda y^{\Delta^j}(\sigma(T)), \quad j \in \{3, \ldots, n-1\}, \tag{2.16}$$
$$y^{\Delta^2}(\sigma(T)) = 0,$$
$$Cy(t_0) + Dy(\sigma(T)) = 0,$$

where C and D are real constants, λ is a real constant such that $\lambda \neq 1$, $f \cdot [t_0, T] \times \mathbb{R} \ldots \times \mathbb{R} \to \mathbb{R}$ is a given function and the following condition

$$\det \begin{pmatrix} C+D & Dh_1(\sigma(T), t_0) & Dh_2(\sigma(T), t_0) & Dh_3(\sigma(T), t_0) & \cdots & Dh_{n-2}(\sigma(T), t_0) \\ 0 & 0 & 1 & h_1(\sigma(T), t_0) & \cdots & h_{n-2}(\sigma(T), t_0) \\ 0 & 0 & \lambda-1 & \lambda h_1(\sigma(T), t_0) & \cdots & \lambda h_{n-5}(\sigma(T), t_0) \\ \vdots & \vdots & \vdots & \vdots & \vdots & \vdots \\ 0 & 0 & 0 & 0 & \cdots & \lambda-1 \end{pmatrix} = 0. \tag{2.17}$$

Let

$$A(t) = c_0 + c_1 h_1(t, t_0) + \cdots + c_{n-2} h_{n-2}(t, t_0), \quad t \in [t_0, T],$$

where c_j, $j \in \{0, \ldots, n-2\}$, are real constants such that

$$A^{\Delta^j}(t_0) = \lambda A^{\Delta^j}(\sigma(T)), \quad j \in \{3, \ldots, n-1\},$$
$$A^{\Delta^2}(\sigma(T)) = 0, \tag{2.18}$$
$$CA(t_0) + DA(\sigma(T)) = 0.$$

We have

$$A(t_0) = c_0,$$
$$A(\sigma(T)) = c_0 + c_1 h_1(\sigma(T), t_0) + \cdots + c_{n-2} h_{n-2}(\sigma(T), t_0)$$

and

$$A^{\Delta}(t) = c_1 + c_2 h_1(t, t_0) + \cdots + c_{n-2} h_{n-3}(t, t_0),$$
$$A^{\Delta^2}(t) = c_2 + c_3 h_1(t, t_0) + \cdots + c_{n-2} h_{n-4}(t, t_0),$$
$$A^{\Delta^j}(t) = c_j + \sum_{k=j+1}^{n} c_k h_{k-j}(t, t_0), \quad t \in [t_0, T], \quad j \in \{3, \ldots, n-1\}.$$

Hence, applying (2.18), we find the following system

$$Cc_0 + D(c_0 + c_1 h_1(\sigma(T), t_0) + \cdots$$
$$+ c_{n-2} h_{n-2}(\sigma(T), t_0)) = 0$$
$$c_2 + c_3 h_1(\sigma(T), t_0) + \cdots$$
$$+ c_{n-2} h_{n-4}(\sigma(T), t_0) = 0$$
$$c_j = \lambda c_j + \lambda \sum_{k=j+1}^{n} c_k h_{k-j}(\sigma(T), t_0),$$
$$j \in \{3, \ldots, n-1\},$$

or

$$(C + D)c_0 + c_1 Dh_1(\sigma(T), t_0) + c_2 Dh_2(\sigma(T), t_0)$$
$$+ c_3 Dh_3(\sigma(T), t_0) + \cdots + c_{n-2} Dh_{n-2}(\sigma(T), t_0) = 0$$
$$c_2 + c_3 h_1(\sigma(T), t_0) + \cdots + c_{n-2} h_{n-4}(\sigma(T), t_0) = 0$$
$$(\lambda - 1)c_j + \lambda \sum_{k=j+1}^{n-1} c_k h_{k-j}(\sigma(T), t_0) = 0,$$
$$j \in \{3, \ldots, n-1\}.$$

By the condition (2.17), it follows that the last system has a nonzero solution. Then, an ANN trial solution of the considered BVP is given by

$$y_a(t, p) = \sum_{j=0}^{n-2} c_j h_j(t, t_0) + h_{n+1}(t, t_0)h_{n+1}(t, \sigma(T))N(t, p), \quad t \in [t_0, T],$$

where $N(t, p)$ is the output of the ANN with input t and parameters p.

2.4   Computation of Gradient for Multilinear ANN Model

Suppose that $\mathbb{T}_r, \mathbb{T}_{uj}, \mathbb{T}_{wjr}, \mathbb{T}_{vj}, j \in \{1, \ldots, m\}, r \in \{1, \ldots, n\}$, are time scales with forward jump operators and delta differentiation operators $\sigma_r, \sigma_{uj}, \sigma_{wjr}, \sigma_{vj}, j \in \{1, \ldots, m\}, r \in \{1, \ldots, n\}$, and $\Delta_r, \Delta_{uj}, \Delta_{wjr}, \Delta_{vj}, j \in \{1, \ldots, m\}, r \in \{1, \ldots, n\}$, respectively.

Consider the network output $N(t, p)$ given by

$$N(t, p) = \sum_{j=1}^{m} v_j s(w_j t + u_j), \tag{2.19}$$

where

$$w_j = (w_{j1}, \ldots, w_{jn}) \quad \text{and} \quad t = \begin{pmatrix} t_1 \\ \vdots \\ t_n \end{pmatrix}.$$

The function (2.19) can be rewritten in the following form

$$N(t, p) = \sum_{j=1}^{m} v_j s \left(\sum_{k=1}^{n} w_{jk} t_k + u_j \right).$$

We will find the partial derivatives of N with respect to v_j, w_{jk}, u_j and t_k, $j \in \{1, \ldots, m\}, k \in \{1, \ldots, n\}$.

1. Assume that $u_j = \mathbb{T}_{vj}, w_{jk} = \mathbb{T}_{wjk}, u_j = \mathbb{T}_{uj}, j = \{1, \ldots, m\}, k = \{1, \ldots, n\}$. For this aim we have the following cases.

   a. Let $t_k \mathbb{T}_{wjk} + u_j, j \in \{1, \ldots, m\}, k \in \{1, \ldots, n\}$, be a time scale with forward jump operator and delta differentiation operator $\tilde{\sigma}_{wjk}$ and $\tilde{\Delta}_{wjk}, j \in \{1, \ldots, m\}, k \in \{1, \ldots, n\}$, respectively, such that

   $$t_k \sigma_{wjk}(w_{jk}) + u_j = \tilde{\sigma}_{wjk}(t_k w_{jk} + u_j), \quad j \in \{1, \ldots, m\}, \quad k \in \{1, \ldots, n\}.$$

   Then

   $$\frac{\partial N}{\Delta w_{lr}}(t, p) = v_l s^{\tilde{\Delta}_{wlr}} \left(\sum_{k=1}^{n} w_{lk} t_k + u_l \right) t_r, \quad l \in \{1, \ldots, m\}, \quad r \in \{1, \ldots, n\}.$$

b. Assume that $t_k \mathbb{T} w_{jk} + u_j$, $j \in \{1, \ldots, m\}$, $k \in \{1, \ldots, n\}$, is not any time scale. Then, using the Pötzsche chain rule, we obtain

$$
\frac{\partial N}{\Delta w_{lr}}(t, .p) = v_l \left(\int_0^1 s' \left(h \sigma_{w_{lr}}(w_{lr}) t_r + (1 - h) w_{lr} t_r + \sum_{k=1, k \neq r}^{n} w_{lk} t_k + u_l \right) dh \right) t_r,
$$

$l \in \{1, \ldots, m\}$, $r \in \{1, \ldots, n\}$.

c. Let $w_{lk} \mathbb{T}_k + u_l$, $l \in \{1, \ldots, m\}$, $k \in \{1, \ldots, n\}$, be a time scale with forward jump operator and delta differentiation operator $\tilde{\sigma}_{lk}$ and $\tilde{\Delta}_{lk}$, $j \in \{1, \ldots, m\}$, $k \in \{1, \ldots, n\}$, respectively, such that

$$
w_{lk} \sigma_k(t_k) + u_l = \tilde{\sigma}_{lk}(t_k w_{lk} + u_l), \quad l \in \{1, \ldots, m\}, \quad k \in \{1, \ldots, n\}.
$$

Then

$$
\frac{\partial N}{\Delta t_r}(t, p) = \sum_{j=1}^{m} v_j s^{\tilde{\Delta}_{lk}} \left(\sum_{k=1}^{n} w_{jk} t_k + u_j \right) w_{jr}, \quad l \in \{1, \ldots, m\}, \quad r \in \{1, \ldots, n\}.
$$

d. Assume that $w_{lk} \mathbb{T}_k + u_l$, $l \in \{1, \ldots, m\}$, $k \in \{1, \ldots, n\}$, is not any time scale. Then, using the Pötzsche chain rule, we obtain

$$
\frac{\partial N}{\Delta t_r}(t, p) = \sum_{j=1}^{m} v_j \left(\int_0^1 s' \left(h w_{jr} \sigma_r(t_r) + (1 - h) w_{jr} t_r + \sum_{k=1, k \neq r}^{n} w_{jk} t_k + u_j \right) dh \right) w_{jr},
$$

$l \in \{1, \ldots, m\}$, $r \in \{1, \ldots, n\}$.

e. Let $\sum_{k=1}^{n} w_{jk} t_k + \mathbb{T}_{uj}$, $j \in \{1, \ldots, n\}$, be a time scale with forward jump operator and delta differentiation operator $\tilde{\sigma}_{uj}$ and $\tilde{\Delta}_{uj}$, $j \in \{1, \ldots, m\}$, $k \in \{1, \ldots, n\}$, respectively, such that

$$
\sum_{k=1}^{n} w_{jk} t_k + \sigma_{uj}(u_j) = \tilde{\sigma}_{uj}(t_k w_{jk} + u_j), \quad j \in \{1, \ldots, m\}.
$$

Then

$$
\frac{\partial N}{\Delta u_r}(t, p) = v_r s^{\tilde{\Delta}_{ur}} \left(\sum_{k=1}^{n} w_{rk} t_k + u_r \right), \quad r \in \{1, \ldots, m\}.
$$

f. Assume that $t_k \mathbb{T} w_{jk} + u_j$, $j \in \{1, \ldots, m\}$, $k \in \{1, \ldots, n\}$, is not any time scale. Then, using the Pötzsche chain rule, we obtain

$$\frac{\partial N}{\Delta u_r}(t, p) = v_r \left(\int_0^1 s' \left(h\sigma_{ur}(u_r) + (1 - h)u_r + \sum_{k=1}^n w_{rk}t_k \right) dh \right),$$

$r \in \{1, \ldots, m\}$.

2. Let v_j, $j \in \{1, \ldots, n\}$, does not belong to any time scale. Then

$$\frac{\partial N}{\partial v_r}(t, p) = s \left(\sum_{k=1}^n w_{rk}t_k + u_r \right), \quad r \in \{1, \ldots, m\}.$$

3. Let w_{jk}, $j \in \{1, \ldots, m\}$, $k \in \{1, \ldots, n\}$, does not belong to any time scale. Then

$$\frac{\partial N}{\partial w_{rl}}(t, p) = v_r s' \left(\sum_{k=1}^n w_{rk}t_k + u_r \right) t_l, \quad r \in \{1, \ldots, m\}, \quad l \in \{1, \ldots, n\}.$$

4. Let u_j, $j \in \{1, \ldots, n\}$, does not belong to any time scale. Then

$$\frac{\partial N}{\partial u_r}(t, p) = v_r s' \left(\sum_{k=1}^n w_{rk}t_k + u_r \right), \quad r \in \{1, \ldots, m\}.$$

2.5 Examples

In this section, we will illustrate the results in the previous sections with some examples.

Example 2.2 Let $\mathbb{T} = \mathbb{Z}$. Consider the following IVP

$$y^\Delta = t + 1, \quad t \in [0, 5],$$
$$y(0) = 0.$$

We will show that

$$y(t) = \frac{1}{2}(t^2 + t), \quad t \in [0, 5],$$

is the exact solution of the considered equation. We have

$$\sigma(t) = t + 1,$$
$$\mu(t) = 1, \quad t \in [0, 5],$$

and

$$y^{\Delta}(t) = \frac{1}{2}(\sigma(t) + t + 1)$$
$$= \frac{1}{2}(t + 1 + t + 1)$$
$$= t + 1, \quad t \in [0, 5],$$

and

$$y(0) = 0.$$

Let the activation function be the time scale sigmoid function for $\mathbb{T} = \mathbb{Z}$, i.e., the function

$$f(t) = \frac{1}{1 + \left(\frac{1}{2}\right)^t}, \quad t \in \mathbb{T}.$$

According to the results in Sect. 2.2.1, we have that the ANN trial solution of the considered IVP is given by the expression

$$y_a(t, p) = tN(t, p), \quad t \in [0, 5].$$

The network data are as follows:

1. the training points

$$t_1 = 1,$$
$$t_2 = 2,$$
$$t_3 = 3,$$
$$t_4 = 4,$$
$$t_5 = 5.$$

2. the input weights

$$w_1 = 1,$$
$$w_2 = 2,$$
$$w_3 = 1,$$
$$w_4 = 3,$$
$$w_5 = 1.$$

3. the hidden weights

$$v_1 = 2,$$
$$v_2 = 1,$$
$$v_3 = 1,$$
$$v_4 = 2,$$
$$v_5 = 2.$$

4. the biases

$$u_1 = 1,$$
$$u_2 = 2,$$
$$u_3 = 1,$$
$$u_4 = 2,$$
$$u_5 = 1.$$

5. the network rate

$$\eta = 0.001.$$

In Fig. 2.1 are given two network data updates. In Fig. 2.2 are given the first three errors. In Fig. 2.3 they are shown the desired result and the first three iterations. In Fig. 2.4 they are shown the first three errors.

Example 2.3 Let $\mathbb{T} = 2^{\mathbb{N}_0}$. Consider the following IVP

$$y^\Delta = 3t + 1, \quad t \in [1, 32],$$
$$y(1) = 2.$$

We will show that

$$y(t) = t^2 + t, \quad t \in [1, 32],$$

is the exact solution of the considered equation. We have

$$\sigma(t) = 2t,$$
$$\mu(t) = t, \quad t \in [1, 32],$$

and

$$y^\Delta(t) = \sigma(t) + t + 1$$
$$= 2t + t + 1$$
$$= 3t + 1, \quad t \in [1, 32],$$

Column1	Column2	Column3	Column4	Column5	Column6	Column7
training points t		desired results y		input weights w		hidden weight
1		1		1		2
2		3		2		1
3		6		1		1
4		10		3		2
5		15		1		2
weights w^1		weights v^1		biases u^1		output results
0.9994		1.9994		0.9994		1.59925384
2.00403077		1.00403077		2.00403077		0.9887111
1.01517647		1.01517647		1.01517647		0.95778171
3.03200049		2.03200049		2.03200049		2.03188949
1.06515385		2.06515385		1.06515385		2.0408347
weights w^2		weights v^2		biases u^2		output results
0.99880075		1.99880075		0.99880075		1.59850864
2.00805335		1.00805335		2.00805335		0.99279849
1.03030313		1.03030313		1.03030313		0.97431562
3.06387293		2.06387293		2.06387293		2.06377198
1.12994967		2.12994967		1.12994967		2.11073902

Fig. 2.1 Two network data updates and the corresponding errors for Example 2.2

Column1	Column2	Column3	Column4	Column5
Error E^1		Error E^2		Error E^3
0.18		0.179553		0.179106
2.030888		2.022642		2.014429
12.79585		12.71198		12.62875
32.00098		31.74539		31.49186
84.90047		83.96998		83.06652

Fig. 2.2 The first three errors for Example 2.2

and

$$y(1) = 2.$$

Let the activation function be the time scale sigmoid function for $\mathbb{T} = 2^{\mathbb{N}_0}$, i.e., the function

$$f(z) = \frac{1}{1 + \prod\limits_{\tau=1}^{\frac{t}{2}} \frac{1}{1+\tau z}}, \quad t \in \mathbb{T}.$$

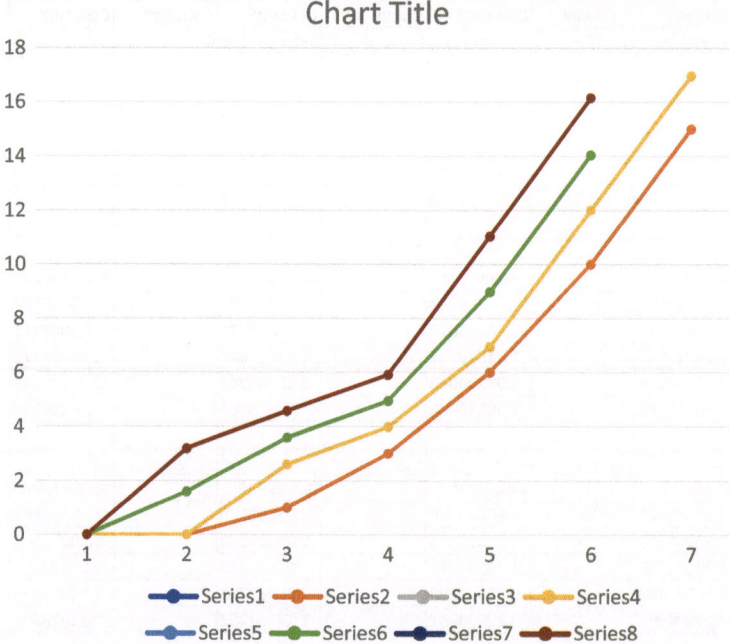

Fig. 2.3 The graphics of the desired result and first three iterations for Example 2.2

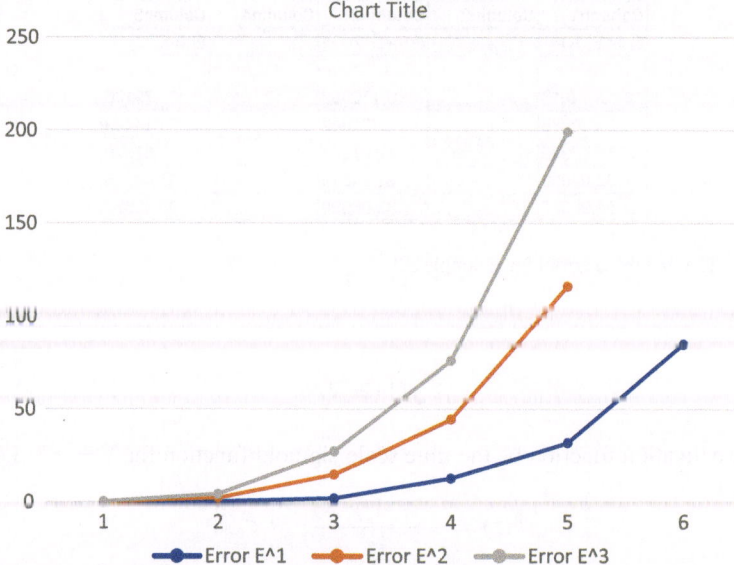

Fig. 2.4 The first three errors for Example 2.2

According to the results in Sect. 2.2.1, we have that the ANN trial solution of the considered IVP is given by the expression

$$y_a(t, p) = 2 + (t - 1)N(t, p), \quad t \in [1, 32].$$

The network data are as follows:

1. the training points

$$t_1 = 2,$$
$$t_2 = 4,$$
$$t_3 = 8,$$
$$t_4 = 16,$$
$$t_5 = 32.$$

2. the input weights

$$w_1 = 1,$$
$$w_2 = 2,$$
$$w_3 = 1,$$
$$w_4 = 3,$$
$$w_5 = 1.$$

3. the hidden weights

$$v_1 = 2,$$
$$v_2 = 1,$$
$$v_3 = 1,$$
$$v_4 = 2,$$
$$v_5 = 2.$$

4. the biases

$$u_1 = 1,$$
$$u_2 = 2,$$
$$u_3 = 1,$$
$$u_4 = 2,$$
$$u_5 = 1.$$

5. the network rate

$$\eta = 0.001.$$

Column1	Column2	Column3	Column4	Column5	Column6	Column7
training points t		desired results y		input weights w		hidden weight
2		6		1		2
4		20		2		1
8		72		1		1
16		272		3		2
32		1056		1		2
weights w^1		weights v^1		biases u^1		output results
1.0088		2.0088		1.0088		1.60915015
2.07604706		1.07604706		2.07604706		1.0639043
1.56800257		1.56800257		1.56800257		1.56769112
7.32000003		6.32000003		6.32000003		6.31999876
34.728		35.728		34.728		35.728
weights w^2		weights v^2		biases u^2		output results
1.0175817		2.0175817		1.0175817		1.61827763
2.15179144		1.15179144		2.15179144		1.13930717
2.13146104		2.13146104		2.13146104		2.13116062
11.57088		10.57088		10.57088		10.5708791
67.376704		68.376704		67.376704		68.376704

Fig. 2.5 Two network data updates and the corresponding errors for Example 2.3

Column1	Column2	Column3	Column4	Column5
Error E^1		Error E^2		Error E^3
9.68		9.639781		9.599745
180.7236		179.2879		177.8629
2520.523		2480.355		2440.827
36450		35292.93		34172.59
555458		520477.5		487699.9

Fig. 2.6 The first three errors for Example 2.3

In Fig. 2.5 are given two network data updates. In Fig. 2.6 are given the first three errors. In Fig. 2.7 they are shown the desired result and the first three iterations. In Fig. 2.8 they are shown the first three errors.

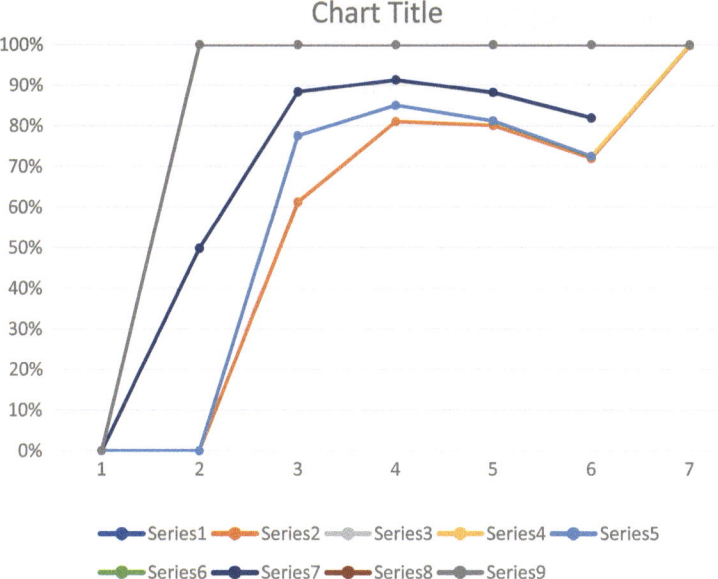

Fig. 2.7 The graphics of the desired result and first three iterations for Example 2.3

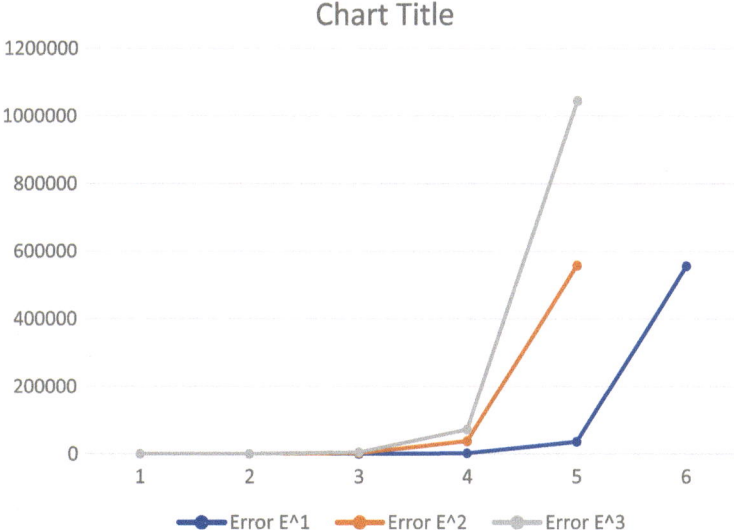

Fig. 2.8 The first three errors for Example 2.3

2.6   Advanced Practical Problems

Problem 2.1 Find an *ANN* trial solution for the following m-point BVP

$$y^{\Delta} = f(t, y), \quad t \in [t_0, T],$$

$$y(t_1) = \sum_{k=2}^{m-1} \alpha_k y(t_k) + \alpha_1 y(\sigma(t_m)),$$

where α_k, $k \in \{1, \ldots, m-1\}$, are given real constants, $f : [t_0, T] \times \mathbb{R} \to \mathbb{R}$ is a given function, and

$$t_0 = t_1 < t_2 < \cdots < t_m = T.$$

Answer 2.3

$$y_a(t, p) = \sum_{k=1}^{m-1} c_k h_1(t, t_k) + c_m h_1(t, \sigma(t_m))$$

$$+ \prod_{k=1}^{m-1} h_2(t, t_k) h_2(t, \sigma(t_m)) N(t, p), \quad t \in [t_0, T],$$

where $c_1, \ldots, c_m$ are real constants that satisfy the system

$$\sum_{k=2}^{m-1} c_k h_1(t_1, t_k) + c_m h_1(t_1, \sigma(t_m))$$

$$= \sum_{k=2}^{m-1} \alpha_k \left(\sum_{l=1, l \neq k}^{m-1} c_l h_1(t_l, t_k) + c_m h_1(t_k, \sigma(t_m)) \right)$$

$$+ \alpha_1 \sum_{k=1}^{m-1} c_k h_1(\sigma(t_m), t_k),$$

$N(t, p)$ is the output of the ANN with input t and parameters p.

Problem 2.2 Find an ANN trial solution to the following BVP

$$y^{\Delta^2} = f(t, y, y^{\Delta}), \quad t \in [t_0, T],$$

$$y(t_0) - y^{\Delta}(t_0) = 0,$$

where $f : [t_0, T] \times \mathbb{R} \times \mathbb{R} \to \mathbb{R}$ is a given function.

Answer 2.4

$$y_a(t, p) = c_2 + c_2 h_1(t, t_0) - c_2 \frac{1}{h_1(\sigma(T), t_0)} h_2(t, t_0)$$
$$+ h_2(t, t_0) h_1(t, \sigma^2(t_0)) h_2(t, \sigma(T)) N(t, p), \quad t \in [t_0, T],$$

where c_2 is an arbitrary real constant and $N(t, p)$ is the output of the ANN with input t and parameters p.

Problem 2.3 Find an ANN trial solution to the following BVP

$$y^{\Delta^2} = f(t, y, y^\Delta), \quad t \in [t_0, T],$$
$$y(t_0) = y(\sigma^2(T)),$$

where $f : [t_0, T] \times \mathbb{R} \times \mathbb{R} \to \mathbb{R}$ is a given function.

Answer 2.5

$$y_a(t, p) = c_1 - \frac{c_1}{h_2(\sigma^2(T), t_0)} h_1(t, t_0) - \frac{c_1}{h_2(t_0, \sigma^2(T))} h_2(t, \sigma^2(T))$$
$$+ h_3(t, t_0) h_3(t, \sigma^2(T)) N(t, p), \quad t \in [t_0, T],$$

where c_1 is an arbitrary real constant and $N(t, p)$ is the output of the ANN with input t and parameters p.

Problem 2.4 Find an ANN trial solution to the following BVP

$$y^{\Delta^2} = f(t, y, y^\Delta), \quad t \in [t_0, T],$$
$$y(t_0) = y^\Delta(\sigma(T)),$$

where $f : [t_0, T] \times \mathbb{R} \times \mathbb{R} \to \mathbb{R}$ is a given function.

Answer 2.6

$$y_a(t, p) = c_1 - \frac{c_1}{h_2(\sigma(T), t_0)} h_1(t, t_0) - \frac{c_1}{h_2(t_0, \sigma(T))} h_2(t, \sigma(T))$$
$$+ h_3(t, t_0) h_3(t, \sigma(T)) N(t, p), \quad t \in [t_0, T],$$

where c_1 is an arbitrary real constant and $N(t, p)$ is the output of the ANN with input t and parameters p.

Problem 2.5 Find an ANN trial solution to the following BVP

$$y^{\Delta^2} = f(t, y, y^\Delta), \quad t \in [t_0, T],$$
$$y(t_0) = ky(\eta),$$
$$y^\Delta(\sigma(T)) = 0,$$

where $f : [t_0, T] \times \mathbb{R} \times \mathbb{R} \to \mathbb{R}$ is a given function, $\eta \in (t_0, T0$ and $k \in \mathbb{R}$ are such that

$$h_2(t_0, \sigma(T)) - kh_2(\eta, \sigma(T)) \neq 0.$$

Answer 2.7

$$
\begin{aligned}
y_a(t, p) = & c_1 h_1(t, t_0) - c_1 h_1(t, \eta) \\
& + \frac{(k-1)c_1 h_1(\eta, t_0)}{h_2(t_0, \sigma(T)) - kh_2(\eta, \sigma(T))} h_2(t, \sigma(T)) \\
& + h_2(t, t_0) h_2(t, \eta) h_3(t, \sigma(T)) N(t, p), \quad t \in [t_0, T],
\end{aligned}
$$

where c_1 is an arbitrary real constant and $N(t, p)$ is the output of the ANN with input t and parameters p.

Problem 2.6 Find an ANN trial solution to the following BVP

$$
\begin{aligned}
y^{\Delta^2} &= f(t, y, y^{\Delta}), \quad t \in [t_0, T], \\
y(t_0) &= A, \\
y(\sigma^2(T)) &= B,
\end{aligned}
$$

where $f : [t_0, T] \times \mathbb{R} \times \mathbb{R} \to \mathbb{R}$ is a given function, A and B are given real constants.

Answer 2.8

$$
\begin{aligned}
y_a(t, p) = & \frac{B}{h_1(\sigma^2(T), t_0)} h_1(t, t_0) + \frac{A}{h_1(t_0, \sigma^2(T))} h_1(t, \sigma^2(T)) \\
& + h_3(t, t_0) h_3(t, \sigma^2(T)) N(t, p), \quad t \in [t_0, T],
\end{aligned}
$$

where c_1 is an arbitrary real constant and $N(t, p)$ is the output of the ANN with input t and parameters p.

Problem 2.7 Find an ANN trial solution to the following BVP

$$
\begin{aligned}
y^{\Delta^3} &= f(t, y, y^{\Delta}, y^{\Delta^2}), \quad t \in [t_0, T], \\
y(t_0) &= 0, \\
y(\sigma(T)) &= 0,
\end{aligned}
$$

where $f : [t_0, T] \times \mathbb{R} \times \mathbb{R} \times \mathbb{R} \to \mathbb{R}$ is a given function.

Answer 2.9

$$y_a(t, p) = -c_4 \left(\frac{h_3(t_0, \sigma(T)) - h_2(t_0, \sigma(T))h_2(\sigma(T), t_0)}{(h_1(t_0, \sigma(T)) + h_2(\sigma(T), t_0))(1 - h_1(t_0, \sigma(T)))} + h_3(t_0, \sigma(T)) \right)$$

$$+ c_4 \frac{1}{h_2(\sigma(T), t_0)}$$

$$\times \left(\frac{h_3(t_0, \sigma(T)) - h_2(t_0, \sigma(T))h_2(\sigma(T), t_0)}{(h_1(t_0, \sigma(T)) + h_2(\sigma(T), t_0))(1 - h_1(t_0, \sigma(T)))} + h_3(t_0, \sigma(T)) \right) h_2(t, t_0)$$

$$+ c_4 \frac{h_3(t_0, \sigma(T)) - h_2(t_0, \sigma(T))h_2(\sigma(T), t_0)}{(h_1(t_0, \sigma(T)) + h_2(\sigma(T), t_0))(1 - h_1(t_0, \sigma(T)))} \frac{h_2(t, \sigma(T))}{h_2(t_0, \sigma(T))}$$

$$+ c_4 h_3(t, \sigma(T))$$

$$+ h_3(t, t_0)h_3(t, \sigma(T))N(t, p), \quad t \in [t_0, T],$$

where c_4 is an arbitrary real constant and $N(t, p)$ is the output of the ANN with input t and parameters p.

Chapter 3
Regression-Based Artificial Neural Networks

In this chapter, the regression-based neural network (RBNN) model has been introduced for solving dynamic equations on arbitrary time scales. The RBNN trial solution of dynamic equations has been obtained by using the RBNN model for single input and single output system. The number of nodes in the hidden layer has been fixed according to the degree of the polynomial in regression fitting, and the coefficients involved are taken as initial weights to start with neural training. Here, the unsupervised error back-propagation method has been used for minimizing the error function. Modifications of the parameters are done without the use of any optimization technique. Initial weights from input to hidden and from hidden to output layer are taken as combination of random (arbitrary) as well as by RBNN. In this chapter, a variety of initial and boundary value problems have been solved.

Suppose that $\mathbb{T}$ is a time scale with forward jump operator and delta differentiation operator σ and Δ, respectively. Let $t_0, T \in \mathbb{T}$ be such that $t_0 < T$.

3.1 Algorithm of RBNN Model

Consider the following patterns as

$$(t_1, y_1), \quad (t_2, y_2), \quad \ldots, (t_n, y_n),$$

where $t_j \in \mathbb{T}$, $y_j \in \mathbb{R}$, $j \in \{1, \ldots, n\}$, the initial weights $w_{ik} \in \mathbb{R}$, $j \in \{1, \ldots, n\}$, $k \in \{0, 1, \ldots, m\}$, the hidden weights $v_{jk} \in \mathbb{R}$, $j \in \{1, \ldots, n\}$, $k \in \{0, 1, \ldots, m\}$. For any value t_j, $j \in \{1, \ldots, n\}$, we may find y_j, $j \in \{1, \ldots, n\}$, by other numerical methods. These methods are usually iterative in nature. After the solution is obtained, if we want to know the solution in between steps, then again we have to iterate the procedure from the beginning. ANN may be one of the ways where we may overcome their prediction of iterations.

© The Author(s), under exclusive license to Springer Nature Switzerland AG 2025   51
S. Georgiev, *Neural Network Methods for Dynamic Equations on Time Scales*,
SpringerBriefs in Computational Intelligence,
https://doi.org/10.1007/978-3-031-85056-1_3

Consider the polynomial

$$p(t_j) = w_{j0} + w_{j1}h_1(t_j, t_0) + w_{j2}h_2(t_j, t_0) + \cdots + w_{jm}h_m(t_j, t_0), \quad j \in \{1, \ldots, n\}. \tag{3.1}$$

Then, we calculate

$$h_0^j = \frac{1}{1 + e_{\ominus(\lambda w_{j0})}(t_j, t_0)},$$

$$h_1^j = \frac{1}{1 + e_{\ominus(\lambda w_{j1}h_1)}(t_j, t_0)},$$

$$h_2^j = \frac{1}{1 + e_{\ominus(\lambda w_{j2}h_2)}(t_j, t_0)},$$

$$\vdots$$

$$h_m^j = \frac{1}{1 + e_{\ominus(\lambda w_{jm}h_m)}(t_j, t_0)}, \quad j \in \{1, \ldots, n\}.$$

The regression analysis is applied to find the output of the network by the relation

$$y^1(t_j) = v_{j0}h_0^j + v_{j1}h_1^j + v_{j2}h_2^j + \cdots + v_{jm}h_m^j, \quad j \in \{1, \ldots, n\}. \tag{3.2}$$

The error is given by the expression

$$E = \sum_{j=1}^{n} \frac{1}{2}(y_j - y^1(t_j))^2.$$

If the error is large, we update the network data as follows

$$w^{1jk} = w_{jk} + \eta(y_j - y^1(t_j))t_j,$$
$$v^{1jk} = v_{jk} + \eta(y_j - y^1(t_j))t_j, \quad j \in \{1, \ldots, n\}, \quad k \in \{0, 1, \ldots, m\}.$$

We continue this process while we minimize the error.

Example 3.1 Let $\mathbb{T} = \mathbb{Z}$. Assume that the input is

$$t = 2,$$

the desired result is

$$y_1 = 2,$$

the initial weights are

$$w_0 = 2,$$
$$w_1 = 2,$$

$$w_2 = 1,$$

the hidden weights are

$$v_0 = 2,$$
$$v_1 = 1,$$
$$v_2 = 1$$

and the network rate is

$$\eta = 0.001.$$

Here

$$\sigma(t) = t + 1,$$
$$\mu(t) = 1, \quad t \in \mathbb{T},$$

and

$$h_1(t, 0) = t,$$
$$h_2(t, 0) = \frac{1}{2}t^2 - \frac{1}{2}t, \quad t \in \mathbb{T}.$$

Take $\lambda = 1$. We have

$$e_{\ominus w_0}(2, 0) = \frac{1}{(1 + w_0)^2},$$
$$e_{\ominus(w_1 h_1)}(2, 0) = \frac{1}{1 + w_1},$$
$$e_{\ominus(w_2 h_2)}(2, 0) = 1.$$

The neural results are given in Fig. 3.1.

Example 3.2 Let $\mathbb{T} = 2^{\mathbb{N}_0}$. Assume that the input is

$$t = 4,$$

the desired result is

$$y_1 = 2,$$

the initial weights are

$$w_0 = 1,$$
$$w_1 = 2,$$
$$w_2 = 3,$$

Column1	Column2	Column3	Column4	Column5	Column6	Column7
t	y_1	w_0	w_1	w_2	v_0	v_1
2	2	2	2	1	2	1
h_0^1	h_1^1	h_2^1	y^1	E^1		
0.9	0.75	0.5	3.05	0.55125		
w_0^1	w_1^1	w_2^1	v_0^1	v_1^1	v_2^1	
2.002205	2.002205	1.002205	1.9979	0.9979	0.9979	
h_0^2	h_1^2	h_2^2	y^2	E^2		
0.90013217	0.75013774	0.5	3.04588652	0.5469393		
w_0^2	w_1^2	w_2^2	v_0^2	v_1^2	v_2^2	
2.00011323	2.00011323	1.00011323	1.99580823	0.99580823	0.99580823	
h_0^3	h_1^3	h_2^3	y^3	E^3		
0.90000679	0.75000708	0.5	3.04100829	0.54184913		

Fig. 3.1 The neural results for Example 3.1

the hidden weights are

$$v_0 = 2,$$
$$v_1 = 3,$$
$$v_2 = 1$$

and the network rate is

$$\eta = 0.001.$$

Here

$$\sigma(t) = 2t,$$
$$\mu(t) = t, \quad t \in \mathbb{T},$$

and

$$h_1(t, 1) = t - 1,$$
$$h_2(t, 1) = \frac{1}{3}t^2 - t + \frac{2}{3}, \quad t \in \mathbb{T}.$$

Column1	Column2	Column3	Column4	Column5	Column6	Column7
t	y_1	w_0	w_1	w_2	v_0	v_1
4	2	1	2	3	2	3
h_0^1	h_1^1	h_2^1	y^1	E^1		
0.85714286	0.83333333	0.5	4.71428571	3.68367347		
w_0^1	w_1^1	w_2^1	v_0^1	v_1^1	v_2^1	
1.02946939	2.02946939	3.02946939	1.98914286	2.98914286	0.98914286	
h_0^2	h_1^2	h_2^2	y^2	E^2		
0.8612657	0.8349546	0.5	4.7035505	3.65459266		
w_0^2	w_1^2	w_2^2	v_0^2	v_1^2	v_2^2	
1.01865519	2.01865519	3.01865519	1.97832866	2.97832866	0.97832866	
h_0^3	h_1^3	h_2^3	y^3	E^3		
0.85977277	0.83436333	0.5	4.67508563	3.57804157		

Fig. 3.2 The neural results for Example 3.2

Take $\lambda = 1$. We have

$$e_{\ominus w_0}(4, 1) = \frac{1}{(1 + w_0)(1 + 2w_0)},$$

$$e_{\ominus(w_1 h_1)}(4, 1) = \frac{1}{1 + 2w_1},$$

$$e_{\ominus(w_2 h_2)}(4, 1) = 1.$$

The neural results are given in Fig. 3.2.

3.2   Learning Algorithm of RBNN Model

The RBNN trial solution $y_a(t, p)$ for dynamic equations on arbitrary time scales with network parameters p (weights, biases) may be written as follows

$$y_a(t, p) = A(t) + F(t, N(t, p)), \quad t \in \mathbb{T}.$$

The first term $A(t)$ does not contain adjustable parameters and satisfies only the initial and/or boundary conditions. The second term $F(t, N(t, p))$ contains the single output $N(t, p)$ of RBNN with input t and adjustable parameters p.

Consider a three-layer network with one input node t, $m + 1$ initial weights w_j, $j \in \{0, 1, \ldots, m\}$, and one hidden layer consisting of $m + 1$ number of nodes v_j, $j \in \{0, 1, \ldots, m\}$. Then the output is defined in the following manner

$$N(t, p) = \sum_{j=0}^{m} v_j h^j(t)$$

$$= \sum_{j=0}^{m} v_j \frac{1}{1 + e_{\ominus(\lambda w_j h_j)}(t, t_0)}.$$

Training the neural network means updating the parameters so that the error values converge to the desired accuracy.

Note that

$$N_t^\Delta(t, p) = \left(\sum_{j=0}^{m} v_j \frac{1}{1 + e_{\ominus(\lambda w_j h_j)}(\cdot, t_0)} \right)^\Delta (t)$$

$$= \sum_{j=0}^{m} v_j \left(\frac{1}{1 + e_{\ominus(\lambda w_j h_j)}(\cdot, t_0)} \right)^\Delta (t)$$

$$= -\sum_{j=0}^{m} v_j \frac{\ominus(\lambda w_j h_j(\cdot, t_0))(t) e_{\ominus(\lambda w_j h_j)}(t, t_0)}{(1 + e_{\ominus(\lambda w_j h_j)}(t, t_0))(1 + e_{\ominus(\lambda w_j h_j)}(\sigma(t), t_0))}.$$

3.3 Formulation for Initial Value Problems

In this section, we will use some computations from Chap. 2.

1. Consider the IVP (2.2). Then the RBNN trial solution is given by the expression

$$y_a(t, p) = y_0 + h_1(t, t_0) \left(\sum_{j=0}^{m} v_j \frac{1}{1 + e_{\ominus(\lambda w_j h_j)}(t, t_0)} \right), \quad t \in [t_0, T].$$

2. Consider the IVP (2.4). Then the RBNN trial solution is given by the expression

$$y_a(t, p) = y_0 + y_1 h_1(t, t_0) + h_2(t, t_0) \left(\sum_{j=0}^{m} v_j \frac{1}{1 + e_{\ominus(\lambda w_j h_j)}(t, t_0)} \right), \quad t \in [t_0, T].$$

3. Consider the IVP (2.5). Then the RBNN trial solution is given by the expression

$$y_a(t, p) = \sum_{j=0}^{n-1} y_j h_j(t, t_0) + h_n(t, t_0) \left(\sum_{j=0}^{m} v_j \frac{1}{1 + e_{\ominus(\lambda w_j h_j)}(t, t_0)} \right), \quad t \in [t_0, T].$$

3.4 Formulation for Boundary Value Problems

In this section, we will use some computations in Chap. 2.

1. Consider the BVP (2.7). Then the RBNN trial solution is given by the expression

$$y_a(t, p) = C + \frac{D - C}{h_1(T, t_0)} h_1(t, t_0)$$

$$+ h_2(t, t_0)h_2(t, T) \left(\sum_{j=0}^{m} v_j \frac{1}{1 + e_{\ominus(\lambda w_j h_j)}(t, t_0)} \right) \quad t \in [t_0, T].$$

2. Consider the BVP (2.8). Then the RBNN trial solution is given by the expression

$$y_a(t, p) = c_1 - \frac{(M + R)c_1}{Rh_1(\sigma(T), t_0)} h_1(t, t_0)$$

$$+ h_2(t, t_0)h_2(t, \sigma(T)) \left(\sum_{j=0}^{m} v_j \frac{1}{1 + e_{\ominus(\lambda w_j h_j)}(t, t_0)} \right), \quad t \in [t_0, T],$$

where c_1 is an arbitrary real constant.

Exercise 3.1 Find the RBNN trial solution for the BVP in Exercise 2.1.

Answer 3.1

$$y_a(t, p) = c + h_2(t, t_0)h_2(t, \sigma(T)) \left(\sum_{j=0}^{m} v_j \frac{1}{1 + e_{\ominus(\lambda w_j h_j)}(t, t_0)} \right), \quad t \in [t_0, T],$$

where c is an arbitrary real constant.

3. Consider the BVP (2.9). Then the RBNN trial solution is given by the expression

$$y_a(t, p) = C + Dh_1(t, t_0) + h_2(t, t_0)h_2(t, T) \left(\sum_{j=0}^{m} v_j \frac{1}{1 + e_{\ominus(\lambda w_j h_j)}(t, t_0)} \right), \quad t \in [t_0, T].$$

4. Consider the BVP (2.10). Then the RBNN trial solution is given by the expression

$$y_a(t, p) = \frac{\beta}{\alpha} c_2 + c_2 h_1(t, t_0)$$

$$- c_2 \frac{\gamma\beta + \alpha\gamma h_1(\sigma^2(t_0), t_0) + \alpha\delta}{\alpha(\gamma h_2(\sigma^2(t_0), t_0) + \delta h_1(\sigma(T), t_0))} h_2(t, t_0)$$

$$+ h_2(t, t_0)h_1(t, \sigma^2(t_0))h_2(t, \sigma(T)) \left(\sum_{j=0}^{m} v_j \frac{1}{1 + e_{\ominus(\lambda w_j h_j)}(t, t_0)} \right), \quad t \in [t_0, T],$$

where c_2 is an arbitrary real constant.

Exercise 3.2   Find the RBNN trial solution for the BVP in Exercise 2.2

Answer 3.2

$$y_a(t, p) = c_2 h_1(t, t_0) - c_2 \frac{1}{h_1(\sigma(T), t_0)} h_2(t, t_0)$$

$$+ h_2(t, t_0) h_1(t, \sigma^2(t_0)) h_2(t, \sigma(T)) \left(\sum_{j=0}^{m} v_j \frac{1}{1 + e_{\ominus(\lambda w_j h_j)}(t, t_0)} \right), \quad t \in [t_0, T],$$

where c_2 is an arbitrary real constant.

6. Consider the BVP (2.16). Then the RBNN trial solution is given by the expression

$$y_a(t, p) = \sum_{j=0}^{n-2} c_j h_j(t, t_0) + h_n(t, t_0) h_n(t, \sigma(T)) \left(\sum_{j=0}^{m} v_j \frac{1}{1 + e_{\ominus(\lambda w_j h_j)}(t, t_0)} \right),$$

$$t \in [t_0, T].$$

3.5   Examples

In this section, we will illustrate the main results in this chapter with several examples.

Example 3.3   Let $\mathbb{T} = \mathbb{Z}$. Consider the following IVP

$$y^{\Delta^2} = 6t + 8, \quad t \in [0, 10],$$
$$y(0) = 0,$$
$$y^{\Delta}(0) = 1.$$

Let the initial input, initial weights, hidden weights and network rate be as in Example 3.1. We will show that

$$y(t) = t^3 + t^2 + t, \quad t \in [0, 10],$$

is the solution of the considered problem. We have

$$\sigma(t) = t + 1,$$
$$\mu(t) = 1, \quad t \in \mathbb{T},$$

and $Ry(0) = 0$, and

$$y^{\Delta}(t) = (\sigma(t))^2 + t\sigma(t) + t^2 + \sigma(t) + t - 1$$

$$= (t+1)^2 + t(t+1) + t^2 + t + 1 + t - 1$$
$$= t^2 + 2t + 1 + t^2 + t + t^2 + 2t$$
$$= 3t^2 + 5t + 1, \quad t \in [0, 10],$$

and $y^\Delta(0) = 1$, and

$$y^{\Delta^2}(t) = 3(\sigma(t) + t) + 5$$
$$= 3(t + 1 + t) + 5$$
$$= 3(2t + 1) + 5$$
$$= 6t + 3 + 5$$
$$= 6t + 8, \quad t \in [0, 10].$$

The neural results are given in Fig. 3.3.

Example 3.4 Let $\mathbb{T} = 2^{\mathbb{N}_0}$. Consider the following IVP

$$y^{\Delta^2} = 21t - 6, \quad t \in \mathbb{T},$$
$$y(1) = -1,$$
$$y^\Delta(1) = 2.$$

Assume that the input, initial weights, hidden weights and network rate are the same as in Example 3.2. We will show that the function

Column1	Column2	Column3	Column4	Column5	Column6	Column7
t	y_1	w_0	w_1	w_2	v_0	v_1
2	10	2	2	1	2	1
h_0^1	h_1^1	h_2^1	y^1	E^1		
0.9	0.75	0.5	5.05	12.25125		
w_0^1	w_1^1	w_2^1	v_0^1	v_1^1	v_2^1	
2.049005	2.049005	1.049005	2.0099	1.0099	1.0099	
h_0^2	h_1^2	h_2^2	y^2	E^2		
0.90287897	0.75302574	0.5	5.08012715	12.1025744		
w_0^2	w_1^2	w_2^2	v_0^2	v_1^2	v_2^2	
2.05884475	2.05884475	1.05884475	2.01973975	1.01973975	1.01973975	
h_0^3	h_1^3	h_2^3	y^3	E^3		
0.90344257	0.83971384	0.5	5.19087832	11.5638257		

Fig. 3.3 The neural results for Example 3.3

$$y(t) = t^3 - 2t^2 + t - 1, \quad t \in \mathbb{T},$$

is the solution of the considered IVP. We have

$$\sigma(t) = 2t,$$
$$\mu(t) = t, \quad t \in \mathbb{T},$$

and

$$y(1) = 1 - 2 + 1 - 1$$
$$= -1,$$

and

$$y^{\Delta}(t) = (\sigma(t))^2 + t\sigma(t) + t^2 - 2(\sigma(t) + t) + 1$$
$$= (2t)^2 + 2t^2 + t^2 - 2(2t + t) + 1$$
$$= 7t^2 - 6t + 1, \quad t \in \mathbb{T},$$

and

$$y^{\Delta}(1) = 7 - 6 + 1$$
$$= 2,$$

and

$$y^{\Delta^2}(t) = 7(\sigma(t) + t) - 6$$
$$= 7(2t + t) - 6$$
$$= 21t - 6, \quad t \in \mathbb{T}.$$

Note that

$$y(4) = 4^4 - 2 \cdot 4^4 + 4 - 1$$
$$= 64 - 32 + 3$$
$$= 35,$$

which is the desired output. The neural results are given in Fig. 3.4.

Example 3.5 Let $\mathbb{T} = \mathbb{Z}$. Consider the following BVP

$$y^{\Delta^2} = y^2 - t^4 + 14t^3 - 59t^2 + 70t - 23, \quad t \in [0, 5],$$
$$y(0) - y^{\Delta}(5) = 0.$$

Here

Column1	Column2	Column3	Column4	Column5	Column6	Column7	Column8
t	y_1	w_0	w_1	w_2	v_0	v_1	v_2
4	35	1	2	3	2	3	1
h_0^1	h_1^1	h_2^1	y^1	E^1			
0.857143	0.833333	0.5	38	4.5			
w_0^1	w_1^1	w_2^1	v_0^1	v_1^1	v_2^1		
1.036	2.036	3.036	1.988	2.988	0.988		
h_0^2	h_1^2	h_2^2	y^2	E^2			
0.862156	0.83531	0.5	37.9271	4.283965			
w_0^2	w_1^2	w_2^2	v_0^2	v_1^2	v_2^2		
1.024292	2.024292	3.024292	1.976292	2.976292	2.976292		
h_0^3	h_1^3	h_2^3	y^3	E^3			
0.860554	0.834672	0.5	36.20193	0.722318			

Fig. 3.4 The neural results for Example 3.4

$$\sigma(t) = t + 1,$$
$$\mu(t) = 1, \quad t \in \mathbb{T}.$$

We will prove that

$$y(t) = t^2 - 7t + 5, \quad t \in [0, 5],$$

is a solution of the considered BVP. We have

$$y(0) = 5$$

and

$$y^{\Delta}(t) = \sigma(t) + t - 7$$
$$= t + 1 + t - 7$$
$$= 2t - 6, \quad t \in \mathbb{T}.$$

Hence,

$$y^{\Delta}(5) = 2 \cdot 5 - 6$$
$$= 10 - 6$$
$$= 4$$

and
$$y^{\Delta^2}(t) = 2, \quad t \in \mathbb{T}.$$

Then

$$
\begin{aligned}
(y(t))^2 &- t^4 + 14t^3 - 59t^2 + 70t - 23 \\
&= (t^2 - 7t + 5)^2 - t^4 + 14t^3 - 59t^2 + 70t - 23 \\
&= t^4 + 49t^2 + 25 - 14t^3 + 10t^2 - 70t - t^4 + 14t^3 - 59t^2 + 70t - 23 \\
&= 2 \\
&= y^{\Delta^2}(t), \quad t \in \mathbb{T}.
\end{aligned}
$$

Next,

$$
\begin{aligned}
y(0) - y^{\Delta}(5) &= 5 - 4 \\
&= 1
\end{aligned}
$$

and

$$
\begin{aligned}
y(2) &= 2^2 - 7 \cdot 2 + 5 \\
&= 4 - 14 + 5 \\
&= -5.
\end{aligned}
$$

Now, assume that the input, initial weights, hidden weights and network rate are the same as in Example 3.1. Let the desired input be -5. The neural results are given in Fig. 3.5.

Example 3.6 Let $\mathbb{T} = 2^{\mathbb{N}_0}$. Consider the following BVP

$$y^{\Delta^2} = \frac{1}{10 + y} - \frac{1}{t^3 - 3t^2 + 4t + 11} + 21t - 9, \quad t \in [1, 8],$$
$$y(1) + y^{\Delta}(8) = 383$$

Here

$$
\begin{aligned}
\sigma(t) &= 2t, \\
\mu(t) &= t, \quad t \in \mathbb{T}.
\end{aligned}
$$

We will prove that
$$y(t) = t^3 - 3t^2 + 4t + 1, \quad t \in \mathbb{T},$$

is a solution to the considered BVP. We have

$$y^{\Delta}(t) = (\sigma(t))^2 + t\sigma(t) + t^2 - 3(\sigma(t) + t) + 4$$

Column1	Column2	Column3	Column4	Column5	Column6	Column7
t	y_1	w_0	w_1	w_2	v_0	v_1
2	-5	2	2	1	2	1
h_0^1	h_1^1	h_2^1	y^1	E^1		
0.9	0.75	0.5	5.05	50.50125		
w_0^1	w_1^1	w_2^1	v_0^1	v_1^1	v_2^1	
2.202005	2.202005	1.202005	1.9799	0.9799	0.9799	
h_0^2	h_1^2	h_2^2	y^2	E^2		
0.91113351	0.76201837	0.5	5.04060504	50.4068748		
w_0^2	w_1^2	w_2^2	v_0^2	v_1^2	v_2^2	
2.18192379	2.18192379	1.18192379	1.95981879	0.95981879	0.95981879	
h_0^3	h_1^3	h_2^3	y^3	E^3		
0.91010944	0.7608756	0.5	4.99386168	49.9386357		

Fig. 3.5 The neural results for Example 3.5

$$= (2t)^2 + 2t^2 + t^2 - 3(2t + t) + 4$$
$$= 4t^2 + 3t^2 - 9t + 4$$
$$= 7t^2 - 9t + 4, \quad t \in \mathbb{T}.$$

Hence,

$$y^{\Delta}(8) = 7 \cdot 8^2 - 9 \cdot 8 + 4$$
$$= 7 \cdot 64 - 72 + 4$$
$$= 448 - 72 + 4$$
$$= 448 - 68$$
$$= 380$$

and

$$y^{\Delta^2}(t) - 7(\sigma(t) + t) - 9$$
$$= 7(2t + t) - 9$$
$$= 21t - 9, \quad t \in \mathbb{T}.$$

Next,

$$\frac{1}{10 + y(t)} - \frac{1}{t^3 - 3t^2 + 4t + 11} + 21t - 9$$

$$= \frac{1}{10 + t^3 - 3t^2 + 4t + 1} - \frac{1}{t^3 - 3t^2 + 4t + 11} + 21t - 9$$
$$= 21t - 9$$
$$= y^{\Delta^2}(t), \quad t \in \mathbb{T}.$$

Moreover,

$$y(1) = 1^3 - 3 \cdot 1^2 + 4 \cdot 1 + 1$$
$$= 1 - 3 + 4 + 1$$
$$= 6 - 3$$
$$= 3$$

and

$$y(1) + y^{\Delta}(8) = 3 + 380$$
$$= 383,$$

and

$$y(4) = 4^3 - 3 \cdot 4^2 + 4 \cdot 4 + 1$$
$$= 64 - 48 + 16 + 1$$
$$= 81 - 48$$
$$= 33.$$

Assume that the input, initial weights, hidden weights and network rate are the same as in Example 3.2. Let the desired input be 33. The neural results are given in Fig. 3.6.

3.6 Advanced Practical Problems

Problem 3.1 Find a RBNN trial solution for the m-point BVP in Problem 2.1.

Answer 3.3

$$y_a(t, p) = \sum_{k=1}^{m-1} c_k h_1(t, t_k) + c_m h_1(t, \sigma(t_m))$$

$$+ \prod_{k=1}^{m-1} h_2(t, t_k) h_2(t, \sigma(t_m)) \left(\sum_{j=0}^{m} v_j \frac{1}{1 + e_{\Theta(\lambda w_j h_j)}(t, t_0)} \right), \quad t \in [t_0, T],$$

where $c_1, \ldots, c_m$ are real constants that satisfy the system

Column1	Column2	Column3	Column4	Column5	Column6	Column7	Column8
t	y_1	w_0	w_1	w_2	v_0	v_1	v_2
4	33	1	2	3	2	3	1
h_0^1	h_1^1	h_2^1	y^1	E^1			
0.857143	0.833333	0.5	38	12.5			
w_0^1	w_1^1	w_2^1	v_0^1	v_1^1	v_2^1		
1.1	2.1	3.1	1.98	2.98	0.98		
h_0^2	h_1^2	h_2^2	y^2	E^2			
0.870466	0.83871	0.5	37.99015	12.45078			
w_0^2	w_1^2	w_2^2	v_0^2	v_1^2	v_2^2		
1.080039	2.080039	3.080039	1.960039	2.960039	0.960039		
h_0^3	h_1^3	h_2^3	y^3	E^3			
0.867953	0.837664	0.5	37.62534	10.69687			

Fig. 3.6 The neural results for Example 3.6

$$\sum_{k=2}^{m-1} c_k h_1(t_1, t_k) + c_m h_1(t_1, \sigma(t_m))$$

$$= \sum_{k=2}^{m-1} \alpha_k \left(\sum_{l=1, l \neq k}^{m-1} c_l h_1(t_l, t_k) + c_m h_1(t_k, \sigma(t_m)) \right)$$

$$+ \alpha_1 \sum_{k=1}^{m-1} c_k h_1(\sigma(t_m), t_k).$$

Problem 3.2 Find a RBNN trial solution to the BVP in Problem 2.2.

Answer 3.4

$$y_a(t, p) = c_2 + c_2 h_1(t, t_0) - c_2 \frac{1}{h_1(\sigma(T), t_0)} h_2(t, t_0)$$

$$+ h_2(t, t_0) h_1(t, \sigma^2(t_0)) h_2(t, \sigma(T)) \left(\sum_{j=0}^{m} v_j \frac{1}{1 + e_{\ominus(\lambda w_j h_j)}(t, t_0)} \right), \quad t \in [t_0, T],$$

where c_2 is an arbitrary real constant.

Problem 3.3 Find a RBNN trial solution to the BVP in Problem 2.3.

Answer 3.5

$$
y_a(t, p) = c_1 - \frac{c_1}{h_2(\sigma^2(T), t_0)} h_1(t, t_0) - \frac{c_1}{h_2(t_0, \sigma^2(T))} h_2(t, \sigma^2(T))
$$
$$
+ h_3(t, t_0) h_3(t, \sigma^2(T)) \left(\sum_{j=0}^{m} v_j \frac{1}{1 + e_{\ominus(\lambda w_j h_j)}(t, t_0)} \right), \quad t \in [t_0, T],
$$

where c_1 is an arbitrary real constant.

Problem 3.4 Find a RBNN trial solution to the BVP in Problem 2.4.

Answer 3.6

$$
y_a(t, p) = c_1 - \frac{c_1}{h_2(\sigma(T), t_0)} h_1(t, t_0) - \frac{c_1}{h_2(t_0, \sigma(T))} h_2(t, \sigma(T))
$$
$$
+ h_3(t, t_0) h_3(t, \sigma(T)) \left(\sum_{j=0}^{m} v_j \frac{1}{1 + e_{\ominus(\lambda w_j h_j)}(t, t_0)} \right), \quad t \in [t_0, T],
$$

where c_1 is an arbitrary real constant.

Problem 3.5 Find a RBNN trial solution to the BVP in Problem 2.5.

Answer 3.7

$$
y_a(t, p) = c_1 h_1(t, t_0) - c_1 h_1(t, \eta)
$$
$$
+ \frac{(k - 1) c_1 h_1(\eta, t_0)}{h_2(t_0, \sigma(T)) - k h_2(\eta, \sigma(T))} h_2(t, \sigma(T))
$$
$$
+ h_2(t, t_0) h_2(t, \eta) h_3(t, \sigma(T)) \left(\sum_{j=0}^{m} v_j \frac{1}{1 + e_{\ominus(\lambda w_j h_j)}(t, t_0)} \right), \quad t \in [t_0, T],
$$

where c_1 is an arbitrary real constant.

Problem 3.6 Find a RBNN trial solution to the BVP in Problem 2.6.

Answer 3.8

$$
y_a(t, p) = \frac{B}{h_1(\sigma^2(T), t_0)} h_1(t, t_0) + \frac{A}{h_1(t_0, \sigma^2(T))} h_1(t, \sigma^2(T))
$$
$$
+ h_3(t, t_0) h_3(t, \sigma^2(T)) \left(\sum_{j=0}^{m} v_j \frac{1}{1 + e_{\ominus(\lambda w_j h_j)}(t, t_0)} \right), \quad t \in [t_0, T],
$$

where c_1 is an arbitrary real constant.

Problem 3.7 Find a RBNN trial solution to the BVP in Problem 2.7.

Answer 3.9

$$
\begin{aligned}
y_a(t, p) =\ & -c_4 \left(\frac{h_3(t_0, \sigma(T)) - h_2(t_0, \sigma(T))h_2(\sigma(T), t_0)}{(h_1(t_0, \sigma(T)) + h_2(\sigma(T), t_0))(1 - h_1(t_0, \sigma(T)))} + h_3(t_0, \sigma(T)) \right) \\
& + c_4 \frac{1}{h_2(\sigma(T), t_0)} \\
& \times \left(\frac{h_3(t_0, \sigma(T)) - h_2(t_0, \sigma(T))h_2(\sigma(T), t_0)}{(h_1(t_0, \sigma(T)) + h_2(\sigma(T), t_0))(1 - h_1(t_0, \sigma(T)))} + h_3(t_0, \sigma(T)) \right) h_2(t, t_0) \\
& + c_4 \frac{h_3(t_0, \sigma(T)) - h_2(t_0, \sigma(T))h_2(\sigma(T), t_0)}{(h_1(t_0, \sigma(T)) + h_2(\sigma(T), t_0))(1 - h_1(t_0, \sigma(T)))} \frac{h_2(t, \sigma(T))}{h_2(t_0, \sigma(T))} \\
& + c_4 h_3(t, \sigma(T)) \\
& + h_3(t, t_0) h_3(t, \sigma(T)) \left(\sum_{j=0}^{m} v_j \frac{1}{1 + e_{\ominus(\lambda w_j h_j)}(t, t_0)} \right), \quad t \in [t_0, T],
\end{aligned}
$$

where c_4 is an arbitrary real constant.

Chapter 4
Chebyshev Neural Networks

In this chapter, the Chebyshev neural network (ChNN) model has been developed for solving dynamic equations on arbitrary time scales. The ChNN trial solution of dynamic equations has been obtained by using the ChNN model for single input and single output system. Initial value problems and boundary value problems have been solved.

Suppose that $\mathbb{T}$ is a time scale with forward jump operator and delta differentiation operator σ and Δ, respectively. In addition, assume that $t_0, T \in \mathbb{T}$ are such that $t_0 < T$.

4.1 Chebyshev Polynomials on Time Scales

There are several kinds Chebyshev polynomials. In particular, we shall introduce the first and second kind polynomials $T_n(t, t_0)$ and $U_n(t, t_0)$, as well as a pair of related Jacobi polynomials $V_n(t, t_0)$ and $W_n(t, t_0)$, which we call Chebyshev polynomials of the third and fourth kind.

Definition 4.1 The Chebyshev polynomials $T_n(t)$ of the first kind are defined as follows

$$T_0(t, t_0) = 1,$$
$$T_1(t, t_0) = h_1(t, t_0),$$
$$T_n(t, t_0) = 2h_1(t, t_0)T_{n-1}(t, t_0) - T_{n-2}(t, t_0), \quad t \in \mathbb{T}, \quad n \in \mathbb{N}, \quad n \geq 2.$$

© The Author(s), under exclusive license to Springer Nature Switzerland AG 2025   69
S. Georgiev, *Neural Network Methods for Dynamic Equations on Time Scales*,
SpringerBriefs in Computational Intelligence,
https://doi.org/10.1007/978-3-031-85056-1_4

We have

$$T_1(t, t_0) = t - t_0,$$
$$T_2(t, t_0) = 2(t - t_0)T_1(t, t_0) - T_0(t, t_0)$$
$$= 2(t - t_0)^2 - 1,$$
$$T_3(t, t_0) = 2(t - t_0)T_2(t, t_0) - T_1(t, t_0)$$
$$= 2(t - t_0)\left(2(t - t_0)^2 - 1\right) - (t - t_0)$$
$$= 4(t - t_0)^3 - 3(t - t_0),$$
$$T_4(t, t_0) = 2(t - t_0)T_3(t, t_0) - T_2(t, t_0)$$
$$= 2(t - t_0)\left(4(t - t_0)^3 - 3(t - t_0)\right) - \left(2(t - t_0)^2 - 1\right)$$
$$= 8(t - t_0)^4 - 6(t - t_0)^2 - 2(t - t_0)^2 + 1$$
$$= 8(t - t_0)^4 - 8(t - t_0)^2 + 1, \quad t \in \mathbb{T},$$

and so on. Coefficients of all polynomials $T_n(t, t_0)$ up to degree $n = 21$ are given in the tables below.

n =	0	1	2	3	4	5	6	7	8	9	10
k =0	1	−1	1	−1	1	−1	1	−1	1	−1	1
1		2	−8	18	−32	50	−72	98	−128	162	−200
2			8	−48	160	−400	840	−1568	2688	−4320	6600
3				32	−256	1120	−3584	9408	−21504	44352	−84480
4					128	−1280	6912	−26880	84480	−228096	549120
5						512	−6144	39424	−180224	658944	−2050048
6							2048	−28672	212992	−1118208	4659200
7								8192	−131072	1105920	−6553600
8									32768	−589824	5570560
9										131072	−2621440
10											524288

Coefficients of $(t - t_0)^{2k}$ in $T_{2n}(t, t_0)$.

n =	0	1	2	3	4	5	6	7	8	9	10
k = 0	1	−3	5	−7	9	−11	13	−15	17	−19	21
1		4	−20	56	−120	220	−364	560	−816	1140	−1540
2			16	−112	432	−1232	2912	−6048	11424	−20064	33264
3				64	−576	2816	−9984	28800	−71808	160512	−329472
4					256	−2816	16640	−70400	239360	−695552	1793792
5						1024	−13312	92160	−452608	1770496	−5870592
6							4096	−61440	487424	−2723840	12042240
7								16384	−278528	2490368	−15597568
8									65536	−1245184	12386304
9										262144	−5505024
10											1048576

Coefficients of $(t - t_0)^{2k+}$ in $T_{2n+1}(t, t_0)$.

Definition 4.2 The Chebyshev polynomials $U_n(t, t_0)$ of the second kind are defined as follows

$$U_0(t, t_0) = 1,$$
$$U_1(t, t_0) = 2h_1(t, t_0),$$
$$U_n(t, t_0) = 2h_1(t, t_0)U_{n-1}(t, t_0) - U_{n-2}(t, t_0), \quad t \in \mathbb{T}, \quad n \in \mathbb{N}, \quad n \geq 2.$$

We have

$$
\begin{aligned}
U_1(t, t_0) &= 2(t - t_0), \\
U_2(t, t_0) &= 2(t - t_0)U_1(t, t_0) - U_0(t, t_0) \\
&= 2(t - t_0)(2(t - t_0)) - 1 \\
&= 4(t - t_0)^2 - 1, \\
U_3(t, t_0) &= 2(t - t_0)U_2(t, t_0) - U_1(t, t_0) \\
&= 2(t - t_0)\left(4(t - t_0)^2 - 1\right) - (t - t_0) \\
&= 8(t - t_0)^3 - 2(t - t_0) - (t - t_0) \\
&= 8(t - t_0)^3 - 3(t - t_0), \\
U_4(t, t_0) &= 2(t - t_0)U_3(t, t_0) - U_2(t, t_0) \\
&= 2(t - t_0)\left(8(t - t_0)^3 - 3(t - t_0)\right) - 4(t - t_0)^2 + 1 \\
&= 16(t - t_0)^4 - 6(t - t_0)^2 - 4(t - t_0)^2 + 1 \\
&= 16(t - t_0)^4 - 10(t - t_0)^2 + 1, \quad t \in \mathbb{T},
\end{aligned}
$$

and so on. Coefficients of all polynomials $U_n(t, t_0)$ up to degree $n = 21$ are given in the tables below.

n =	0	1	2	3	4	5	6	7	8	9	10
k = 0	1	-1	1	-1	1	-1	1	-1	1	-1	1
1		4	-12	24	-40	60	-84	112	-144	180	-220
2			16	-80	240	-560	1120	-2016	3360	-5280	7920
3				64	-448	1792	-5376	13440	-29568	59136	-109824
4					256	-2304	11520	-42240	126720	-329472	768768
5						1024	-11264	67584	-292864	1025024	-3075072
6							4096	-53248	372736	-1863680	7454720
7								16384	-245760	1966080	-11141120
8									65536	-1114112	10027008
9										262144	-4980736
10											1048576

Coefficients of $(t - t_0)^{2k}$ in $U_{2k}(t, t_0)$.

n =	0	1	2	3	4	5	6	7	8	9	10
k = 0	2	-4	6	-8	10	-12	14	-16	18	-20	22
1		8	-32	80	-160	280	-448	672	-960	1320	-1760
2			32	-192	672	-1792	4032	-8064	14784	-25344	41184
3				128	-1024	4608	-15360	42240	-101376	219648	-439296
4					512	-5120	28160	-112640	366080	-1025024	2562560
5						2048	-24576	159744	-745472	2795520	-8945664
6							8192	-114688	860160	-4587520	19496960
7								32768	-524288	4456448	-26738688
8									131072	-2359296	22413312
9										524288	-10485760
10											2097152

Coefficients of $(t - t_0)^{2k+1}$ in $U_{2k+1}(t, t_0)$.

The following relation between the Chebyshev polynomials of the first kind and the second kind is valid.

Theorem 4.1 *For any $n \in \mathbb{N}$, $n \geq 2$, one has the following relation*

$$U_n(t, t_0) - U_{n-2}(t, t_0) = 2T_n(t, t_0), \quad t \in \mathbb{T}. \tag{4.1}$$

Proof We will use the principal of the mathematical induction.

1. For $n = 2$, one has

$$\begin{aligned}
U_2(t, t_0) - U_0(t, t_0) &= 2h_1(t, t_0)U_1(t, t_0) - U_0(t, t_0) - U_0(t, t_0) \\
&= 4(h_1(t, t_0))^2 - 2 \\
&= 2\left(2(h_1(t, t_0))^2 - 1\right) \\
&= 2T_2(t, t_0), \quad t \in \mathbb{T}.
\end{aligned}$$

2. Assume that (4.1) holds for some $n \in \mathbb{N}$, $n \geq 2$.
3. We will prove that

$$U_{n+1}(t, t_0) - U_{n-1}(t, t_0) = 2T_{n+1}(t, t_0), \quad t \in \mathbb{T}.$$

Really, using (4.1) and

$$U_{n-1}(t, t_0) - U_{n-3}(t, t_0) = 2T_{n-1}(t, t_0), \quad t \in \mathbb{T},$$

one get

$$\begin{aligned}
U_{n+1}(t, t_0) - U_{n-1}(t, t_0) &= 2h_1(t, t_0)U_n(t, t_0) - U_{n-1}(t, t_0) \\
&\quad -2h_1(t, t_0)U_{n-2}(t, t_0) + U_{n-3}(t, t_0) \\
&= 2h_1(t, t_0)(U_n(t, t_0) - U_{n-2}(t, t_0)) \\
&\quad -(U_{n-1}(t, t_0) - U_{n-3}(t, t_0)) \\
&= 4h_1(t, t_0)T_n(t, t_0) - 2T_{n-1}(t, t_0)
\end{aligned}$$

$$= 2 \left(2h_1(t, t_0) T_n(t, t_0) - T_{n-1}(t, t_0)\right)$$
$$= 2T_{n+1}(t, t_0), \quad t \in \mathbb{T}.$$

Now, applying the principal of the mathematical induction, we conclude that the equality (4.1) is valid for any $n \in \mathbb{N}$, $n \geq 2$. This completes the proof.

Definition 4.3 The Chebyshev polynomials $V_n(t, t_0)$ of the third kind are defined as follows

$$V_0(t, t_0) = 1,$$
$$V_1(t, t_0) = 2h_1(t, t_0) - 1,$$
$$V_n(t, t_0) = 2h_1(t, t_0) V_{n-1}(t, t_0) - V_{n-2}(t, t_0), \quad t \in \mathbb{T}, \quad n \in \mathbb{N}, \quad n \geq 2.$$

We have

$$V_1(t, t_0) = 2(t - t_0) - 1,$$
$$V_2(t, t_0) = 2(t - t_0) V_1(t, t_0) - V_0(t, t_0)$$
$$= 2(t - t_0)(2(t - t_0) - 1) - 1$$
$$= 4(t - t_0)^2 - 2(t - t_0) - 1,$$
$$V_3(t, t_0) = 2(t - t_0) V_2(t, t_0) - V_1(t, t_0)$$
$$= 2(t - t_0) \left(4(t - t_0)^2 - 2(t - t_0) - 1\right) - 2(t - t_0) + 1$$
$$= 8(t - t_0)^3 - 4(t - t_0)^2 - 2(t - t_0) - 2(t - t_0) + 1$$
$$= 8(t - t_0)^3 - 4(t - t_0)^2 - 4(t - t_0) + 1,$$
$$V_4(t, t_0) = 2(t - t_0) V_3(t, t_0) - V_2(t, t_0)$$
$$= 2(t - t_0) \left(8(t - t_0)^3 - 4(t - t_0)^2 - 4(t - t_0) + 1\right)$$
$$\quad -4(t - t_0)^2 + 2(t - t_0) + 1$$
$$= 16(t - t_0)^4 - 8(t - t_0)^3 - 8(t - t_0)^2 + 2(t - t_0)$$
$$\quad -4(t - t_0)^2 + 2(t - t_0) + 1$$
$$= 16(t - t_0)^4 - 8(t - t_0)^3 - 12(t - t_0)^2 + 4(t - t_0) + 1, \quad t \in \mathbb{T},$$

and so on.

Definition 4.4 The Chebyshev polynomials $W_n(t, t_0)$ of the fourth kind are defined as follows

$$W_0(t, t_0) = 1,$$
$$W_1(t, t_0) = 2h_1(t, t_0) + 1,$$
$$W_n(t, t_0) = 2h_1(t, t_0) W_{n-1}(t, t_0) - W_{n-2}(t, t_0), \quad t \in \mathbb{T}, \quad n \in \mathbb{N}, \quad n \geq 2.$$

We have

$$W_1(t, t_0) = 2(t - t_0) + 1,$$
$$W_2(t, t_0) = 2(t - t_0)W_1(t, t_0) - W_0(t, t_0)$$
$$= 2(t - t_0)(2(t - t_0) + 1) - 1$$
$$= 4(t - t_0)^2 + 2(t - t_0) - 1,$$
$$W_3(t, t_0) = 2(t - t_0)W_2(t, t_0) - W_1(t, t_0)$$
$$= 2(t - t_0)\left(4(t - t_0)^2 + 2(t - t_0) - 1\right) - 2(t - t_0) - 1$$
$$= 8(t - t_0)^3 + 4(t - t_0)^2 - 2(t - t_0) - 2(t - t_0) - 1$$
$$= 8(t - t_0)^3 + 4(t - t_0)^2 - 4(t - t_0) - 1,$$
$$W_4(t, t_0) = 2(t - t_0)W_3(t, t_0) - W_2(t, t_0)$$
$$= 2(t - t_0)\left(8(t - t_0)^3 + 4(t - t_0)^2 - 4(t - t_0) - 1\right)$$
$$\qquad - 4(t - t_0)^2 - 2(t - t_0) + 1$$
$$= 16(t - t_0)^4 + 8(t - t_0)^3 - 8(t - t_0)^2 - 2(t - t_0)$$
$$\qquad - 4(t - t_0)^2 - 2(t - t_0) + 1$$
$$= 16(t - t_0)^4 + 8(t - t_0)^3 - 12(t - t_0)^2 - 4(t - t_0) + 1, \quad t \in \mathbb{T},$$

and so on. Below, the coefficients of $(t - t_0)^k$ in $V_n(t, t_0)$ and $(-1)^{n+k}(t - t_0)^k$ in $W_n(t, t_0)$ are given.

n =	0	1	2	3	4	5	6	7	8	9	10
k = 0	1	−1	−1	1	1	−1	−1	1	1	−1	−1
1		2	−2	−4	4	6	−6	−8	8	10	−10
2			4	−4	−12	12	24	−24	−40	40	60
3				8	−8	−32	32	80	−80	−160	160
4					16	−16	−80	80	240	−240	−560
5						32	−32	−192	192	672	−672
6							64	−64	−448	448	1792
7								128	−128	−1024	1024
8									256	−256	−2304
9										512	−512
10											1024

We have the following relation.

Theorem 4.2 *For any $n \in \mathbb{N}_0$, one has*

$$U_n(t, t_0) = \frac{1}{2}(V_n(t, t_0) + W_n(t, t_0)), \quad t \in \mathbb{T}. \tag{4.2}$$

Proof We will use the principal of the mathematical induction.

1. For $n = 0$, one has

$$U_0(t, t_0) = 1$$

$$= \frac{1}{2}(1+1)$$

$$= \frac{1}{2}(V_0(t, t_0) + W_0(t, t_0)).$$

2. Assume that (4.2) holds for some $n \in \mathbb{N}$.
3. We will prove that

$$U_{n+1}(t, t_0) = \frac{1}{2}(V_{n+1}(t, t_0) + W_{n+1}(t, t_0)), \quad t \in \mathbb{T}.$$

Really, using (4.2) and

$$U_{n-1}(t, t_0) = \frac{1}{2}(V_{n-1}(t, t_0) + W_{n-1}(t, t_0)), \quad t \in \mathbb{T},$$

one get

$$
\begin{aligned}
U_{n+1}(t, t_0) &= 2h_1(t, t_0)U_n(t, t_0) - U_{n-1}(t, t_0) \\
&= h_1(t, t_0)(V_n(t, t_0) + W_n(t, t_0)) \\
&\quad -\frac{1}{2}(V_{n-1}(t, t_0) + W_{n-1}(t, t_0)) \\
&= \frac{1}{2}(2h_1(t, t_0)V_n(t, t_0) - V_{n-1}(t, t_0)) \\
&\quad +\frac{1}{2}(2h_1(t, t_0)W_n(t, t_0) - W_{n-1}(t, t_0)) \\
&= \frac{1}{2}V_{n+1}(t, t_0) + \frac{1}{2}W_{n+1}(t, t_0), \quad t \in \mathbb{T}.
\end{aligned}
$$

Applying the principal of the mathematical induction, we conclude that (4.2) holds for any $n \in \mathbb{N}$. This completes the proof.

Theorem 4.3 *For any $n \in \mathbb{N}$, one has*

$$U_{n-1}(t, t_0) = \frac{1}{2}(W_n(t, t_0) - V_n(t, t_0)), \quad t \in \mathbb{T}. \tag{4.3}$$

Proof For the proof we will use the principal of the mathematical induction.

1. Let $n = 1$. Then

$$
\begin{aligned}
\frac{1}{2}(W_1(t, t_0) - V_1(t, t_0)) &= \frac{1}{2}(2h_1(t, t_0) + 1 - 2h_1(t, t_0) + 1) \\
&= 1 \\
&= U_0(t, t_0), \quad t \in \mathbb{T}.
\end{aligned}
$$

Let now, $n = 2$. Then

$$\frac{1}{2}(W_2(t, t_0) - V_2(t, t_0)) = \frac{1}{2}\bigg(2h_1(t, t_0)W_1(t, t_0) - W_0(t, t_0)$$

$$-2h_1(t, t_0)V_1(t, t_0) + V_0(t, t_0)\bigg)$$

$$= h_1(t, t_0)(W_1(t, t_0) - V_1(t, t_0))$$

$$= 2h_1(t, t_0)U_0(t, t_0)$$

$$= 2h_1(t, t_0)$$

$$= U_1(t, t_0), \quad t \in \mathbb{T}.$$

2. Assume that (4.3) holds for some $n \in \mathbb{N}$, $n \geq 3$.
3. We will prove that

$$U_n(t, t_0) = \frac{1}{2}(W_{n+1}(t, t_0) - V_{n+1}(t, t_0)), \quad t \in \mathbb{T}.$$

Really, using (4.3) and

$$U_{n-2}(t, t_0) = \frac{1}{2}(W_{n-1}(t, t_0) - V_{n-1}(t, t_0)), \quad t \in \mathbb{T},$$

one get

$$\frac{1}{2}(W_{n+1}(t, t_0) - V_{n+1}(t, t_0)) = \frac{1}{2}(2h_1(t, t_0)W_n(t, t_0) - W_{n-1}(t, t_0)$$

$$-2h_1(t, t_0)V_n(t, t_0) + V_{n-1}(t, t_0))$$

$$= \frac{1}{2}(2h_1(t, t_0)(W_n(t, t_0) - V_n(t, t_0))$$

$$-(W_{n-1}(t, t_0) - V_{n-1}(t, t_0)))$$

$$= \frac{1}{2}(4h_1(t, t_0)U_{n-1}(t, t_0) - 2U_{n-2}(t, t_0))$$

$$= 2h_1(t, t_0)U_{n-1}(t, t_0) - U_{n-2}(t, t_0)$$

$$= U_n(t, t_0), \quad t \in \mathbb{T}.$$

Applying the principal of the mathematical induction, we conclude that (4.3) holds for any $n \in \mathbb{N}$. This completes the proof.

Theorem 4.4 *For any $n \in \mathbb{N}$, $n \geq 2$, one has*

$$U_{n-2}(t, t_0) = U_n(t, t_0) - 2T_n(t, t_0), \quad t \in \mathbb{T}. \tag{4.4}$$

Proof We will use the principal of the mathematical induction.

1. Let $n = 2$. Then

$$
\begin{aligned}
U_2(t, t_0) - 2T_2(t, t_0) &= 2h_1(t, t_0)U_1(t, t_0) - U_0(t, t_0) \\
&\quad -4h_1(t, t_0)T_1(t, t_0) + 2T_0(t, t_0) \\
&= 4(h_1(t, t_0))^2 - 1 - 4(h_1(t, t_0))^2 + 2 \\
&= 1 \\
&= U_0(t, t_0), \quad t \in \mathbb{T}.
\end{aligned}
$$

Let now, $n = 3$. Then

$$
\begin{aligned}
U_3(t, t_0) - 2T_3(t, t_0) &= 2h_1(t, t_0)U_2(t, t_0) - U_1(t, t_0) \\
&\quad -4h_1(t, t_0)T_2(t, t_0) + 2T_1(t, t_0) \\
&= 2h_1(t, t_0)(U_2(t, t_0) - 2T_2(t, t_0)) \\
&\quad -2h_1(t, t_0) + 2h_1(t, t_0) \\
&= 2h_1(t, t_0)U_0(t, t_0) \\
&= 2h_1(t, t_0) \\
&= U_1(t, t_0), \quad t \in \mathbb{T}.
\end{aligned}
$$

2. Assume that (4.4) holds for some $n \geq 4$.
3. We will prove that

$$
U_{n-1}(t, t_0) = U_{n+1}(t, t_0) - 2T_{n+1}(t, t_0), \quad t \in \mathbb{T}.
$$

Really, using (4.4) and

$$
U_{n-3}(t, t_0) = U_{n-1}(t, t_0) - 2T_{n-1}(t, t_0), \quad t \in \mathbb{T},
$$

one get

$$
\begin{aligned}
U_{n+1}(t, t_0) - 2T_{n+1}(t, t_0) &= 2h_1(t, t_0)U_n(t, t_0) - U_{n-1}(t, t_0) \\
&\quad -4h_1(t, t_0)T_n(t, t_0) + 2T_{n-1}(t, t_0) \\
&= 2h_1(t, t_0)(U_n(t, t_0) - 2T_n(t, t_0)) \\
&\quad -(U_{n-1}(t, t_0) - 2T_{n-1}(t, t_0)) \\
&= 2h_1(t, t_0)U_{n-2}(t, t_0) - U_{n-3}(t, t_0) \\
&= U_{n-1}(t, t_0), \quad t \in \mathbb{T}.
\end{aligned}
$$

Hence, applying the principal of the mathematical induction, we conclude that (4.4) holds for any $n \in \mathbb{N}, n \geq 2$. This completes the proof.

Theorem 4.5 *For any $n \in \mathbb{N}$, one has*

$$h_1(t, t_0)U_{n-1}(t, t_0) = U_n(t, t_0) - T_n(t, t_0), \quad t \in \mathbb{T}. \tag{4.5}$$

Proof We will use the principal of the mathematical induction.

1. Let $n = 1$. Then

$$\begin{aligned}
U_1(t, t_0) - T_1(t, t_0) &= 2h_1(t, t_0) - h_1(t, t_0) \\
&= h_1(t, t_0) \\
&= h_1(t, t_0)U_0(t, t_0), \quad t \in \mathbb{T}.
\end{aligned}$$

Let now, $n = 2$. Then

$$\begin{aligned}
U_2(t, t_0) - T_2(t, t_0) &= 2h_1(t, t_0)U_1(t, t_0) - U_0(t, t_0) \\
&\quad -2h_1(t, t_0)T_1(t, t_0) + T_0(t, t_0) \\
&= 2h_1(t, t_0)(U_1(t, t_0) - T_1(t, t_0)) - 1 + 1 \\
&= 2(h_1(t, t_0))^2 U_0(t, t_0) \\
&= 2(h_1(t, t_0))^2 \\
&= h_1(t, t_0)(2h_1(t, t_0)) \\
&= h_1(t, t_0)U_1(t, t_0), \quad t \in \mathbb{T}.
\end{aligned}$$

2. Assume that (4.5) holds for some $n \in \mathbb{N}$.
3. We will prove that

$$h_1(t, t_0)U_n(t, t_0) = U_{n+1}(t, t_0) - T_{n+1}(t, t_0), \quad t \in \mathbb{T}.$$

Really, using (4.5) and

$$h_1(t, t_0)U_{n-2}(t, t_0) = U_{n-1}(t, t_0) - T_{n-1}(t, t_0), \quad t \in \mathbb{T},$$

one get

$$\begin{aligned}
U_{n+1}(t, t_0) - T_{n+1}(t, t_0) &= 2h_1(t, t_0)U_n(t, t_0) - U_{n-1}(t, t_0) \\
&\quad -2h_1(t, t_0)T_n(t, t_0) + T_{n-1}(t, t_0) \\
&= 2h_1(t, t_0)(U_n(t, t_0) - T_n(t, t_0)) \\
&\quad -(U_{n-1}(t, t_0) - T_{n-1}(t, t_0)) \\
&= 2(h_1(t, t_0))^2 U_{n-1}(t, t_0) - h_1(t, t_0)U_{n-2}(t, t_0) \\
&= h_1(t, t_0)(2h_1(t, t_0)U_{n-1}(t, t_0) - U_{n-2}(t, t_0)) \\
&= h_1(t, t_0)U_n(t, t_0), \quad t \in \mathbb{T}.
\end{aligned}$$

Hence, applying the principle of the mathematical induction, we conclude that (4.5) holds for any $n \in \mathbb{N}$. This completes the proof.

Exercise 4.1 Let $n \in \mathbb{N}$, $n \geq 2$. Prove that

$$T_n(t, t_0) = h_1(t, t_0)U_{n-1}(t, t_0) - U_{n-2}(t, t_0), \quad t \in \mathbb{T}.$$

Below, for convenience, with $C_n(t, t_0)$ we will denote the Chebyshev polynomial of the first kind or of the second kind, or of the third kind, or of the fourth kind.

4.2 Algorithm of Chebyshev Neural Network Model

In this section, we will describe the algorithm for Chebyshev neural networks (ChNN). Consider the following patterns as

$$(t_1, y_1), \quad (t_2, y_2), \quad \ldots, (t_n, y_n),$$

where $t_j \in \mathbb{T}$, $y_j \in \mathbb{R}$, $j \in \{1, \ldots, n\}$, the initial weights $w_{jk} \in \mathbb{R}$, $j \in \{1, \ldots, n\}$, $k \in \{0, 1, \ldots, m\}$, the hidden weights $v_{jk} \in \mathbb{R}$, $j \in \{1, \ldots, n\}$, $k \in \{0, 1, \ldots, m\}$.

Compute

$$z_j = \sum_{k=1}^{m} w_{jk}C_{k-1}(t_j, t_0), \quad j \in \{1, \ldots, n\},$$

and

$$\tanh_{z_j}(t_j, t_0) = \frac{e_{z_j}(t_j, t_0) - e_{-z_j}(t_j, t_0)}{e_{z_j}(t_j, t_0) + e_{-z_j}(t_j, t_0)}, \quad j \in \{1, \ldots, n\}.$$

Then, the output of the network is given by the relation

$$y^1(t_j) = v_j \tanh_{z_j}(t_j, t_0), \quad j \in \{1, \ldots, n\}.$$

The error is given by the expression

$$E = \sum_{j=1}^{n} \frac{1}{2}(y_j - y^1(t_j))^2,$$

If the error is large, we update the network data as follows

$$w^{1jk} = w_{jk} + \eta(y_j - y^1(t_j))t_j,$$
$$v^{1jk} = v_{jk} + \eta(y_j - y^1(t_j))t_j, \quad j \in \{1, \ldots, n\}, \quad k \in \{0, 1, \ldots, m\}.$$

We continue this process while we minimize the error.

Example 4.1 Let $\mathbb{T} = \mathbb{Z}$, the input

$$t = 2,$$

the desired result

$$y = 7,$$

the initial weights

$$w_0 = 7,$$
$$w_1 = 3,$$
$$w_2 = 4$$

the hidden weight

$$v = 6$$

and the network rate

$$\eta = 0.001.$$

The neural results for the cases $C_n(t, 0) = T_n(t, 0)$, $C_n(t, 0) = U_n(t, 0)$, $C_n(t, 0) = V_n(t, 0)$ and $C_n(t, 0) = W_n(t, 0)$ are given in Figs. 4.1, 4.2, 4.3 and 4.4, respectively.

Column1	Column2	Column3	Column4	Column5	Column6	Column7	Column8
t	y_1	w_0	w_1	w_2	v		
2	7	7	3	4	6		
z(0)	z(1)		y^1	E^1			
3	14		2.372093	10.70876			
w_0^1	w_1^1	w_2^1	v^1				
7.042005	3.042005	4.042005	6.000250				
z^1(0)	z^1(1)		y^2	E^2			
3	14.12851		2.372441	10.70715			
w_0^2	w_1^2	w_2^2	v^2				
7.05209	3.05209	4.05209	6.018511				
z^2(0)	z^2(1)		y^3	E^3			
3	14.15627		2.375386	10.69353			

Fig. 4.1 Two network data updates and the corresponding errors for Example 4.1 in the case $C_n(t, 0) = T_n(t, 0)$

Column1	Column2	Column3	Column4	Column5	Column6	Column7	Column8
t	y_1	w_0	w_1	w_2	v		
2	7	7	3	4	6		
z(0)	z(1)		y^1	E^1			
3	25		2.210526	11.46953			
w_0^1	w_1^1	w_2^1	v^1				
7.045878	3.045878	4.045878	6.009579				
z^1(0)	z^1(1)		y^2	E^2			
3	25.27527		2.211789	11.46348			
w_0^2	w_1^2	w_2^2	v^2				
7.055455	3.055455	4.055455	6.019155				
z^2(0)	z^2(1)		y^3	E^3			
3	25.33273		2.214846	11.44885			

Fig. 4.2 Two network data updates and the corresponding errors for Example 4.1 in the case $C_n(t, 0) = U_n(t, 0)$

Column1	Column2	Column3	Column4	Column5	Column6	Column7	Column8
t	y_1	w_0	w_1	w_2	v		
2	7	7	3	4	6		
z(0)	z(1)		y^1	E^1			
3	14		2.372093	10.70876			
w_0^1	w_1^1	w_2^1	v^1				
7.042835	3.042835	4.042835	6.009256				
z^1(0)	z^1(1)		y^2	E^2			
3	14.12851		2.372441	10.70715			
w_0^2	w_1^2	w_2^2	v^2				
7.05209	3.05209	4.05209	6.018511				
z^2(0)	z^2(1)		y^3	E^3			
3	14.15627		2.375386	10.69353			

Fig. 4.3 Two network data updates and the corresponding errors for Example 4.1 in the case $C_n(t, 0) = V_n(t, 0)$

Column1	Column2	Column3	Column4	Column5	Column6	Column7	Column8
t	y_1	w_0	w_1	w_2	v		
2	7	7	3	4	6		
z(0)	z(1)		y^1	E^1			
3	36		2.146789	11.77683			
w_0^1	w_1^1	w_2^1	v^1				
7.047107	3.047107	4.047107	6.009706				
z^1(0)	z^1(1)		y^2	E^2			
3	36.42397		2.148566	11.76821			
w_0^2	w_1^2	w_2^2	v^2				
7.05681	3.05681	4.05681	6.019409				
z^2(0)	z^2(1)		y^3	E^3			
3	36.51129		2.15169	11.75305			

Fig. 4.4 Two network data updates and the corresponding errors for Example 4.1 in the case $C_n(t, 0) = W_n(t, 0)$

Example 4.2 Let $\mathbb{T} = 2^{\mathbb{N}_0}$, the input

$$t = 4,$$

the desired result

$$y = 15,$$

the initial weights

$$w_0 = 9,$$
$$w_1 = 1,$$
$$w_2 = 5$$

the hidden weight

$$v = 8$$

and the network rate

$$\eta = 0.001.$$

The neural results for the cases $C_n(t, 1) = T_n(t, 1)$, $C_n(t, 1) = U_n(t, 1)$, $C_n(t, 1) = V_n(t, 1)$ and $C_n(t, 1) = W_n(t, 1)$ are given in Figs. 4.5, 4.6, 4.7 and 4.8, respectively.

Column1	Column2	Column3	Column4	Column5	Column6	Column7	Column8
t	y_1	w_0	w_1	w_2	v		
4	15	9	1	5	8		
z(0)	z(1)		y^1	E^1			
4	15		2.247934	81.3076			
w_0^1	w_1^1	w_2^1	v^1				
9.650461	1.650461	5.650461	8.051008				
z^1(0)	z^1(1)		y^2	E^2			
4	16.95138		2.233754	81.48852			
w_0^2	w_1^2	w_2^2	v^2				
9.701526	1.701526	5.701526	8.102073				
z^2(0)	z^2(1)		y^3	E^3			
4	17.10458		2.245944	81.33297			

Fig. 4.5 Two network data updates and the corresponding errors for Example 4.2 in the case $C_n(t, 1) = T_n(t, 1)$

Column1	Column2	Column3	Column4	Column5	Column6	Column7	Column8
t	y_1	w_0	w_1	w_2	v		
4	15	9	1	5	8		
z(0)	z(1)		y^1	E^1			
4	26		2.143541	82.64427			
w_0^1	w_1^1	w_2^1	v^1				
9.661154	1.661154	5.661154	8.051426				
z^1(0)	z^1(1)		y^2	E^2			
4	29.96693		2.138276	82.71198			
w_0^2	w_1^2	w_2^2	v^2				
9.712601	1.712601	5.712601	8.102873				
z^2(0)	z^2(1)		y^3	E^3			
4	30.27561		2.150657	82.55281			

Fig. 4.6 Two network data updates and the corresponding errors for Example 4.2 in the case $C_n(t, 1) = U_n(t, 1)$

Column1	Column2	Column3	Column4	Column5	Column6	Column7	Column8
t	y_1	w_0	w_1	w_2	v		
4	15	9	1	5	8		
z(0)	z(1)		y^1	E^1			
4	15		2.247934	81.3076			
w_0^1	w_1^1	w_2^1	v^1				
9.650461	1.650461	5.650461	8.051008				
z^1(0)	z^1(1)		y^2	E^2			
4	16.95138		2.233754	81.48852			
w_0^2	w_1^2	w_2^2	v^2				
9.701526	1.701526	5.701526	8.102073				
z^2(0)	z^2(1)		y^3	E^3			
4	17.10458		2.245944	81.33297			

Fig. 4.7 Two network data updates and the corresponding errors for Example 4.2 in the case $C_n(t,1) = V_n(t,1)$

Column1	Column2	Column3	Column4	Column5	Column6	Column7	Column8
t	y_1	w_0	w_1	w_2	v		
4	15	9	1	5	8		
z(0)	z(1)		y^1	E^1			
4	37		2.10101	83.19197			
w_0^1	w_1^1	w_2^1	v^1				
9.665536	1.665536	5.665536	8.051596				
z^1(0)	z^1(1)		y^2	E^2			
4	42.90902		2.100407	83.15930			
w_0^2	w_1^2	w_2^2	v^2				
9.717134	1.717134	5.717134	8.103194				
z^2(0)	z^2(1)		y^3	E^3			
4	43.45421		2.112959	83.03792			

Fig. 4.8 Two network data updates and the corresponding errors for Example 4.2 in the case $C_n(t,1) = W_n(t,1)$

4.3 Learning Algorithm of ChNN Model

The ChNN trial solution $y_a(t, p)$ for dynamic equations on arbitrary time scales with network parameters p (weights, biases) is given in the following manner

$$y_a(t, p) = A(t) + F(t, N(t, p)), \quad t \in \mathbb{T},$$

where $A(t)$ does not contain adjustable parameters and satisfies only the initial and/or boundary conditions and the second term $F(t, N(t, p))$ contains the single output $N(t, p)$ of ChNN with input t and adjustable parameters p.

Consider a three-layer network with one input node t, $m + 1$ initial weights w_j, $j \in \{0, 1, \ldots, m\}$, and one hidden layer consisting of $m + 1$ number of nodes v_j, $j \in \{0, 1, \ldots, m\}$. Then the output is defined in the following manner

$$N(t, p) = \sum_{j=1}^{n} v_j \tanh_{z_j}(t_j, t_0)$$

$$= \sum_{j=1}^{n} v_j \frac{e_{z_j}(t_j, t_0) - e_{-z_j}(t_j, t_0)}{e_{z_j}(t_j, t_0) + e_{-z_j}(t_j, t_0)},$$

where

$$z_j = \sum_{k=1}^{m} w_{jk} C_{k-1}(t_j, t_0), \quad j \in \{1, \ldots, n\},$$

and $C_l(\cdot, t_0)$, $l \in \{0, 1, \ldots, m\}$ are the Chebyshev polynomials of the first kind or second kind, or third kind, or fourth kind. Training the neural network means updating the parameters so that the error values converge to the desired accuracy.

Observe that

$$N_t^\Delta(t, p) = \left(\sum_{j=1}^{n} v_j \frac{e_{z_j}(t_j, t_0) - e_{-z_j}(t_j, t_0)}{e_{z_j}(t_j, t_0) + e_{-z_j}(t_j, t_0)} \right)^\Delta (t)$$

$$= \sum_{j=1}^{n} v_j \left(\frac{e_{z_j}(t_j, t_0) - e_{-z_j}(t_j, t_0)}{e_{z_j}(t_j, t_0) + e_{-z_j}(t_j, t_0)} \right)^\Delta (t)$$

$$= \sum_{j=1}^{n} v_j \left(\frac{1}{(e_{z_j}(t_j, t_0) + e_{-z_j}(t_j, t_0))(e_{z_j}(\sigma(t_j), t_0) + e_{-z_j}(\sigma(t_j), t_0))} \right.$$

$$\left. \Big((z_j e_{z_j}(t_j, t_0) + z_j e_{-z_j}(t_j, t_0))(e_{z_j}(t_j, t_0) + e_{-z_j}(t_j, t_0)) \right.$$

$$\left. -(e_{z_j}(t_j, t_0) - e_{-z_j}(t_j, t_0))(z_j e_{z_j}(t_j, t_0) - z_j e_{-z_j}(t_j, t_0)) \Big) \right)$$

$$= \sum_{j=1}^{n} v_j z_j \left(\frac{1}{(e_{z_j}(t_j, t_0) + e_{-z_j}(t_j, t_0))(e_{z_j}(\sigma(t_j), t_0) + e_{-z_j}(\sigma(t_j), t_0))} \right.$$

$$\left. \Big((e_{z_j}(t_j, t_0) + e_{-z_j}(t_j, t_0))^2 - (e_{z_j}(t_j, t_0) - e_{-z_j}(t_j, t_0))^2 \Big) \right)$$

$$= \sum_{j=1}^{n} v_j z_j \frac{4 e_{z_j}(t_j, t_0) e_{-z_j}(t_j, t_0)}{(e_{z_j}(t_j, t_0) + e_{-z_j}(t_j, t_0))(e_{z_j}(\sigma(t_j), t_0) + e_{-z_j}(\sigma(t_j), t_0))}$$

$$= \sum_{j=1}^{n} v_j z_j \frac{4 e_{z_j \oplus (-z_j)}(t_j, t_0)}{(e_{z_j}(t_j, t_0) + e_{-z_j}(t_j, t_0))(e_{z_j}(\sigma(t_j), t_0) + e_{-z_j}(\sigma(t_j), t_0))}.$$

4.4   Formulation for Initial Value Problems

In this section, we will use some computations from Chap. 2.

1. Consider the IVP (2.2). Then the ChNN trial solution is given by the expression

$$y_a(t, p) = y_0 + h_1(t, t_0) \left(\sum_{j=1}^{n} v_j \frac{e_{z_j}(t_j, t_0) - e_{-z_j}(t_j, t_0)}{e_{z_j}(t_j, t_0) + e_{-z_j}(t_j, t_0)} \right), \quad t \in [t_0, T].$$

2. Consider the IVP (2.4). Then the ChNN trial solution is given by the expression

$$y_a(t, p) = y_0 + y_1 h_1(t, t_0) + h_2(t, t_0) \left(\sum_{j=1}^{n} v_j \frac{e_{z_j}(t_j, t_0) - e_{-z_j}(t_j, t_0)}{e_{z_j}(t_j, t_0) + e_{-z_j}(t_j, t_0)} \right), \quad t \in [t_0, T].$$

3. Consider the IVP (2.5). Then the ChNN trial solution is given by the expression

$$y_a(t, p) = \sum_{j=0}^{n-1} y_j h_j(t, t_0) + h_n(t, t_0) \left(\sum_{j=1}^{n} v_j \frac{e_{z_j}(t_j, t_0) - e_{-z_j}(t_j, t_0)}{e_{z_j}(t_j, t_0) + e_{-z_j}(t_j, t_0)} \right), \quad t \in [t_0, T].$$

4.5   Formulation for Boundary Value Problems

In this section, we will use some computations in Chap. 2.

1. Consider the BVP (2.7). Then the ChNN trial solution is given by the expression

$$y_a(t, p) = C + \frac{D - C}{h_1(T, t_0)} h_1(t, t_0)$$

$$+ h_2(t, t_0) h_2(t, T) \left(\sum_{j=1}^{n} v_j \frac{e_{z_j}(t_j, t_0) - e_{-z_j}(t_j, t_0)}{e_{z_j}(t_j, t_0) + e_{-z_j}(t_j, t_0)} \right), \quad t \in [t_0, T].$$

2. Consider the BVP (2.8). Then the ChNN trial solution is given by the expression

$$y_a(t, p) = c_1 - \frac{(M + R)c_1}{Rh_1(\sigma(T), t_0)} h_1(t, t_0) + h_2(t, t_0)h_2(t, \sigma(T))$$

$$\times \left(\sum_{j=1}^{n} v_j \frac{e_{z_j}(t_j, t_0) - e_{-z_j}(t_j, t_0)}{e_{z_j}(t_j, t_0) + e_{-z_j}(t_j, t_0)} \right), \quad t \in [t_0, T],$$

where c_1 is an arbitrary real constant.

Exercise 4.2 Find the ChNN trial solution for the BVP in Exercise 2.1.

Answer 4.6

$$y_a(t, p) = c + h_2(t, t_0)h_2(t, \sigma(T)) \left(\sum_{j=1}^{n} v_j \frac{e_{z_j}(t_j, t_0) - e_{-z_j}(t_j, t_0)}{e_{z_j}(t_j, t_0) + e_{-z_j}(t_j, t_0)} \right), \quad t \in [t_0, T],$$

where c is an arbitrary real constant.

3. Consider the BVP (2.9). Then the ChNN trial solution is given by the expression

$$y_a(t, p) = C + Dh_1(t, t_0) + h_2(t, t_0)h_2(t, T) \left(\sum_{j=1}^{n} v_j \frac{e_{z_j}(t_j, t_0) - e_{-z_j}(t_j, t_0)}{e_{z_j}(t_j, t_0) + e_{-z_j}(t_j, t_0)} \right), \quad t \in [t_0, T].$$

4. Consider the BVP (2.10). Then the ChNN trial solution is given by the expression

$$y_a(t, p) = \frac{\beta}{\alpha} c_2 + c_2 h_1(t, t_0) - c_2 \frac{\gamma\beta + \alpha\gamma h_1(\sigma^2(t_0), t_0) + \alpha\delta}{\alpha(\gamma h_2(\sigma^2(t_0), t_0) + \delta h_1(\sigma(T), t_0))} h_2(t, t_0)$$

$$+ h_2(t, t_0)h_1(t, \sigma^2(t_0))h_2(t, \sigma(T)) \left(\sum_{j=1}^{n} v_j \frac{e_{z_j}(t_j, t_0) - e_{-z_j}(t_j, t_0)}{e_{z_j}(t_j, t_0) + e_{-z_j}(t_j, t_0)} \right),$$

$$t \in [t_0, T],$$

where c_2 is an arbitrary real constant.

Exercise 4.3 Find the ChNN trial solution for the BVP in Exercise 2.2

Answer 4.7

$$y_a(t, p) = c_2 h_1(t, t_0) - c_2 \frac{1}{h_1(\sigma(T), t_0)} h_2(t, t_0) + h_2(t, t_0)h_1(t, \sigma^2(t_0))h_2(t, \sigma(T))$$

$$\times \left(\sum_{j=1}^{n} v_j \frac{e_{z_j}(t_j, t_0) - e_{-z_j}(t_j, t_0)}{e_{z_j}(t_j, t_0) + e_{-z_j}(t_j, t_0)} \right), \quad t \in [t_0, T],$$

where c_2 is an arbitrary real constant.

5. Consider the BVP (2.16). Then the ChNN trial solution is given by the expression

$$y_a(t, p) = \sum_{j=0}^{n-2} c_j h_j(t, t_0) + h_n(t, t_0) h_n(t, \sigma(T)) \left(\sum_{j=1}^{n} v_j \frac{e_{z_j}(t_j, t_0) - e_{-z_j}(t_j, t_0)}{e_{z_j}(t_j, t_0) + e_{-z_j}(t_j, t_0)} \right),$$

$$t \in [t_0, T].$$

4.6   Examples

The results in this chapter we will illustrate with the following examples.

Example 4.3 Let $\mathbb{T} = \mathbb{Z}$. Consider the following IVP

$$y^{\Delta^2} = \frac{1}{1 + y} - \frac{t^3 + 5t^2 + 7t + 1}{(t + 1)(t + 2)(t + 3)}, \quad t \in [0, 5],$$

$$y(0) = 1,$$

$$y^{\Delta}(0) = -\frac{1}{2}.$$

Here

$$\sigma(t) = t + 1,$$
$$\mu(t) = 1, \quad t \in \mathbb{T}.$$

We will prove that the function

$$y(t) = \frac{1}{1 + t}, \quad t \in [0, 5],$$

is a solution of the considered IVP. We have

$$y^{\Delta}(t) = -\frac{1}{(1 + t)(1 + \sigma(t))}$$

$$= -\frac{1}{(1 + t)(1 + t + 1)}$$

$$= -\frac{1}{(1 + t)(2 + t)}, \quad t \in [0, 5],$$

and

$$y^{\Delta^2}(t) = \frac{\sigma(t) + t + 3}{(1 + t)(2 + t)(1 + \sigma(t))(2 + \sigma(t))}$$

$$= \frac{t+1+t+3}{(1+t)(2+t)(1+t+1)(2+t+1)}$$

$$= \frac{2t+4}{(t+1)(t+2)^2(t+3)}$$

$$= \frac{2(t+2)}{(t+1)(t+2)^2(t+3)}$$

$$= \frac{2}{(t+1)(t+2)(t+3)}, \quad t \in [0,5].$$

Then

$$\frac{1}{1+y(t)} - \frac{t^3 + 5t^2 + 7t + 1}{(t+1)(t+2)(t+3)}$$

$$= \frac{1}{1 + \frac{1}{1+t}} - \frac{t^3 + 5t^2 + 7t + 1}{(t+1)(t+2)(t+3)}$$

$$= \frac{t+1}{t+2} - \frac{t^3 + 5t^2 + 7t + 1}{(t+1)(t+2)(t+3)}$$

$$= \frac{(t+1)^2(t+3) - t^3 - 5t^2 - 7t - 1}{(t+1)(t+2)(t+3)}$$

$$= \frac{(t^2 + 2t + 1)(t+3) - t^3 - 5t^2 - 7t - 1}{(t+1)(t+2)(t+3)}$$

$$= \frac{t^3 + 3t^2 + 2t^2 + 6t + t + 3 - t^3 - 5t^2 - 7t - 1}{(t+1)(t+2)(t+3)}$$

$$= \frac{2}{(t+1)(t+2)(t+3)}$$

$$= y^{\Delta^2(t)}, \quad t \in [0,5].$$

Moreover,

$$y(0) = 1,$$
$$y^{\Delta}(0) = -\frac{1}{2}$$

and

$$y(2) = \frac{1}{3}.$$

Let the input, initial weights, the hidden weight and the network rate be as in Example 4.1. Let also, the output be $\frac{1}{3}$. The neural results for the cases $C_n(t,0) = T_n(t,0)$, $C_n(t,0) = U_n(t,0)$, $C_n(t,0) = V_n(t,0)$ and $C_n(t,0) = W_n(t,0)$ are given in Figs. 4.9, 4.10, 4.11 and 4.12, respectively.

Column1	Column2	Column3	Column4	Column5	Column6	Column7	Column8
t	y_1	w_0	w_1	w_2	v		
2	0.333333	7	3	4	6		
z(0)	z(1)		y^1	E^1			
3	14		2.372093	2.078271			
w_0^1	w_1^1	w_2^1	v^1				
7.008313	3.008313	4.008313	5.995922				
z^1(0)	z^1(1)		y^2	E^2			
3	14.02494		2.369835	2.07367			
w_0^2	w_1^2	w_2^2	v^2				
7.00424	3.00424	4.00424	5.991849				
z^2(0)	z^2(1)		y^3	E^3			
3	14.01272		2.368541	2.071036			

Fig. 4.9 Two network data updates and the corresponding errors for Example 4.3 in the case $C_n(t, 0) = T_n(t, 0)$

Column1	Column2	Column3	Column4	Column5	Column6	Column7	Column8
t	y_1	w_0	w_1	w_2	v		
2	0.333333	7	3	4	6		
z(0)	z(1)		y^1	E^1			
3	25		2.210526	1.761927			
w_0^1	w_1^1	w_2^1	v^1				
7.007048	3.007048	4.007048	5.996246				
z^1(0)	z^1(1)		y^2	E^2			
3	25.04229		2.208793	1.768674			
w_0^2	w_1^2	w_2^2	v^2				
7.003297	3.003297	4.003297	5.992495				
z^2(0)	z^2(1)		y^3	E^3			
3	25.01978		2.207597	1.756432			

Fig. 4.10 Two network data updates and the corresponding errors for Example 4.3 in the case $C_n(t, 0) = U_n(t, 0)$

Column1	Column2	Column3	Column4	Column5	Column6	Column7	Column8
t	y_1	w_0	w_1	w_2	v		
2	0.333333	7	3	4	6		
z(0)	z(1)		y^1	E^1			
3	14		2.372093	2.078271			
w_0^1	w_1^1	w_2^1	v^1				
7.008313	3.008313	4.008313	5.995922				
z^1(0)	z^1(1)		y^2	E^2			
3	14.02494		2.369835	2.07367			
w_0^2	w_1^2	w_2^2	v^2				
7.00424	3.00424	4.00424	5.991849				
z^2(0)	z^2(1)		y^3	E^3			
3	14.01272		2.368541	2.071036			

Fig. 4.11 Two network data updates and the corresponding errors for Example 4.3 in the case $C_n(t, 0) = V_n(t, 0)$

Column1	Column2	Column3	Column4	Column5	Column6	Column7	Column8
t	y_1	w_0	w_1	w_2	v		
2	0.333333	7	3	4	6		
z(0)	z(1)		y^1	E^1			
3	36		2.146789	1.644311			
w_0^1	w_1^1	w_2^1	v^1				
7.006577	3.006577	4.006577	5.996373				
z^1(0)	z^1(1)		y^2	E^2			
3	36.0592		2.145253	1.641526			
w_0^2	w_1^2	w_2^2	v^2				
7.002953	3.002953	4.002953	5.992749				
z^2(0)	z^2(1)		y^3	E^3			
3	36.02658		2.144088	1.639415			

Fig. 4.12 Two network data updates and the corresponding errors for Example 4.3 in the case $C_n(t, 0) = W_n(t, 0)$

Example 4.4 Let $\mathbb{T} = 2^{\mathbb{N}_0}$. Consider the following BVP

$$y^{\Delta^2} = \frac{3(t+4)}{8(t+1)(t+2)(t+3)}(y - y^2), \quad t \in [1, 8],$$

$$y(1) - 12y(8) = -\frac{4}{5}.$$

Here

$$\sigma(t) = 2t,$$
$$\mu(t) = t, \quad t \in \mathbb{T}.$$

We will prove that the function

$$y(t) = \frac{1}{t+4}, \quad t \in [1, 8],$$

is a solution of the considered BVP. We have

$$y^{\Delta}(t) = -\frac{1}{(4+t)(4+\sigma(t))}$$

$$= -\frac{1}{(4+t)(4+2t)}$$

$$= -\frac{1}{2(t+2)(t+4)}, \quad t \in [1, 8],$$

and

$$y^{\Delta^2}(t) = \frac{\sigma(t) + t + 6}{2(t+2)(t+4)(\sigma(t)+2)(\sigma(t)+4)}$$

$$= \frac{2t + t + 6}{2(t+2)(t+4)(2t+2)(2t+4)}$$

$$= \frac{3t + 6}{8(t+1)(t+2)^2(t+4)}$$

$$= \frac{3}{8(t+1)(t+2)(t+4)}, \quad t \in [0, 8].$$

Hence,

$$
\frac{3(t+4)}{8(t+1)(t+2)(t+3)}(y(t)-(y(t))^2)
$$
$$
= \frac{3(t+4)}{8(t+1)(t+2)(t+3)}\left(\frac{1}{t+4}-\frac{1}{(t+4)^2}\right)
$$
$$
= \frac{3(t+4)}{8(t+1)(t+2)(t+3)}\left(\frac{t+3}{(t+4)^2}\right)
$$
$$
= \frac{3}{8(t+1)(t+2)(t+4)}
$$
$$
= y^{\Delta^2}(t), \quad t \in [1,8],
$$

and

$$
y(1) - 12y(8) = \frac{1}{5} - 1
$$
$$
= -\frac{4}{5},
$$

and

$$
y(4) = \frac{1}{8}.
$$

Let the input, initial weights, hidden weight and network rake be as in Example 4.2. Let also, the output be $\frac{1}{8}$. The neural results for the cases $C_n(t, 1) = T_n(t, 1)$, $C_n(t, 1) = U_n(t, 1)$, $C_n(t, 1) = V_n(t, 1)$ and $C_n(t, 1) = W_n(t, 1)$ are given in Tables 4.13, 4.14, 4.15 and 4.16, respectively.

4.7 Advanced Practical Problems

Problem 4.1 Let $n \in \mathbb{N}$, $n \geq 2$. Prove that

$$
U_n(t, t_0) = 2h_1(t, t_0)U_{n-1}(t, t_0) - U_{n-2}(t, t_0), \quad t \in \mathbb{T}.
$$

Problem 4.2 Prove that

$$
T_n(t, t_0) = \det \begin{pmatrix} 2h_1(t, t_0) & -1 & 0 & 0 & \cdots & 0 & 0 & 0 \\ -1 & 2h_1(t, t_0) & -1 & 0 & \cdots & 0 & 0 & 0 \\ 0 & -1 & 2h_1(t, t_0) & -1 & \cdots & 0 & 0 & 0 \\ \vdots & \vdots & \vdots & \vdots & \vdots & \vdots & & \vdots \\ 0 & 0 & 0 & 0 & \cdots & -1 & 2h_1(t, t_0) & -1 \\ 0 & 0 & 0 & 0 & \cdots & 0 & -1 & 2h_1(t, t_0) \end{pmatrix}, \quad t \in \mathbb{T},
$$

($n \times n$ determinant).

Column1	Column2	Column3	Column4	Column5	Column6	Column7	Column8
t	y_1	w_0	w_1	w_2	v		
4	0.125	9	1	5	8		
z(0)	z(1)		y^1	E^1			
4	15		2.247934	2.253424			
w_0^1	w_1^1	w_2^1	v^1				
9.018027	1.018027	5.018027	7.991508				
z^1(0)	z^1(1)		y^2	E^2			
4	15.05408		2.244665	2.246491			
w_0^2	w_1^2	w_2^2	v^2				
9.009549	1.009549	5.009549	7.98303				
z^2(0)	z^2(1)		y^3	E^3			
4	15.02865		2.242698	2.242322			

Fig. 4.13 Two network data updates and the corresponding errors for Example 4.4 in the case $C_n(t, 1) = T_n(t, 1)$

Column1	Column2	Column3	Column4	Column5	Column6	Column7	Column8
t	y_1	w_0	w_1	w_2	v		
4	0.125	9	1	5	8		
z(0)	z(1)		y^1	E^1			
4	26		2.143541	2.037253			
w_0^1	w_1^1	w_2^1	v^1				
9.016298	1.016298	5.016298	7.991926				
z^1(0)	z^1(1)		y^2	E^2			
4	26.09779		2.140843	2.031811			
w_0^2	w_1^2	w_2^2	v^2				
9.008235	1.008235	5.008235	7.983862				
z^2(0)	z^2(1)		y^3	E^3			
4	26.04941		2.138946	2.02799			

Fig. 4.14 Two network data updates and the corresponding errors for Example 4.4 in the case $C_n(t, 1) = U_n(t, 1)$

Column1	Column2	Column3	Column4	Column5	Column6	Column7	Column8
t	y_1	w_0	w_1	w_2	v		
4	0.125	9	1	5	8		
z(0)	z(1)		y^1	E^1			
4	15		2.247934	2.253424			
w_0^1	w_1^1	w_2^1	v^1				
9.018027	1.018027	5.018027	7.991508				
z^1(0)	z^1(1)		y^2	E^2			
4	15.05408		2.244665	2.246491			
w_0^2	w_1^2	w_2^2	v^2				
9.009549	1.009549	5.009549	7.98303				
z^2(0)	z^2(1)		y^3	E^3			
4	15.02865		2.242698	2.242322			

Fig. 4.15 Two network data updates and the corresponding errors for Example 4.4 in the case $C_n(t, 1) = V_n(t, 1)$

Column1	Column2	Column3	Column4	Column5	Column6	Column7	Column8
t	y_1	w_0	w_1	w_2	v		
4	0.125	9	1	5	8		
z(0)	z(1)		y^1	E^1			
4	37		2.10101	1.952308			
w_0^1	w_1^1	w_2^1	v^1				
9.015618	1.015618	5.015618	7.992096				
z^1(0)	z^1(1)		y^2	E^2			
4	37.14057		2.098554	1.947457			
w_0^2	w_1^2	w_2^2	v^2				
9.007724	1.007724	5.007724	7.984202				
z^2(0)	z^2(1)		y^3	E^3			
4	37.06952		2.096673	1.943747			

Fig. 4.16 Two network data updates and the corresponding errors for Example 4.4 in the case $C_n(t, 1) = W_n(t, 1)$

Problem 4.3 Find a ChNN trial solution for the m-point BVP in Problem 2.1.

Answer 4.8

$$y_a(t, p) = \sum_{k=1}^{m-1} c_k h_1(t, t_k) + c_m h_1(t, \sigma(t_m)) + \prod_{k=1}^{m-1} h_2(t, t_k) h_2(t, \sigma(t_m))$$

$$\times \left(\sum_{j=1}^{n} v_j \frac{e_{z_j}(t_j, t_0) - e_{-z_j}(t_j, t_0)}{e_{z_j}(t_j, t_0) + e_{-z_j}(t_j, t_0)} \right), \quad t \in [t_0, T],$$

where $c_1, \ldots, c_m$ are real constants that satisfy the system

$$\sum_{k=2}^{m-1} c_k h_1(t_1, t_k) + c_m h_1(t_1, \sigma(t_m)) = \sum_{k=2}^{m-1} \alpha_k \left(\sum_{l=1, l \neq k}^{m-1} c_l h_1(t_l, t_k) + c_m h_1(t_k, \sigma(t_m)) \right)$$

$$+ \alpha_1 \sum_{k=1}^{m-1} c_k h_1(\sigma(t_m), t_k).$$

Problem 4.4 Find a ChNN trial solution to the BVP in Problem 2.2.

Answer 4.9

$$y_a(t, p) = c_2 + c_2 h_1(t, t_0) - c_2 \frac{1}{h_1(\sigma(T), t_0)} h_2(t, t_0) + h_2(t, t_0) h_1(t, \sigma^2(t_0)) h_2(t, \sigma(T))$$

$$\times \left(\sum_{j=1}^{n} v_j \frac{e_{z_j}(t_j, t_0) - e_{-z_j}(t_j, t_0)}{e_{z_j}(t_j, t_0) + e_{-z_j}(t_j, t_0)} \right), \quad t \in [t_0, T],$$

where c_2 is an arbitrary real constant.

Problem 4.5 Find a ChNN trial solution to the BVP in Problem 2.3.

Answer 4.10

$$y_a(t, p) - c_1 - \frac{c_1}{h_2(\sigma^2(T), t_0)} h_1(t, t_0) - \frac{c_1}{h_2(t_0, \sigma^2(T))} h_2(t, \sigma^2(T))$$

$$+ h_3(t, t_0) h_3(t, \sigma^2(T)) \left(\sum_{j=1}^{n} v_j \frac{e_{z_j}(t_j, t_0) - e_{-z_j}(t_j, t_0)}{e_{z_j}(t_j, t_0) + e_{-z_j}(t_j, t_0)} \right), \quad t \in [t_0, T],$$

where c_1 is an arbitrary real constant.

Problem 4.6 Find a ChNN trial solution to the BVP in Problem 2.4.

Answer 4.11

$$y_a(t, p) = c_1 - \frac{c_1}{h_2(\sigma(T), t_0)} h_1(t, t_0) - \frac{c_1}{h_2(t_0, \sigma(T))} h_2(t, \sigma(T))$$

$$+h_3(t, t_0)h_3(t, \sigma(T)) \left(\sum_{j=1}^{n} v_j \frac{e_{z_j}(t_j, t_0) - e_{-z_j}(t_j, t_0)}{e_{z_j}(t_j, t_0) + e_{-z_j}(t_j, t_0)} \right), \quad t \in [t_0, T],$$

where c_1 is an arbitrary real constant.

Problem 4.7 Find a ChNN trial solution to the BVP in Problem 2.5.

Answer 4.12

$$y_a(t, p) = c_1 h_1(t, t_0) - c_1 h_1(t, \eta) + \frac{(k-1)c_1 h_1(\eta, t_0)}{h_2(t_0, \sigma(T)) - k h_2(\eta, \sigma(T))} h_2(t, \sigma(T))$$

$$+h_2(t, t_0)h_2(t, \eta)h_3(t, \sigma(T)) \left(\sum_{j=1}^{n} v_j \frac{e_{z_j}(t_j, t_0) - e_{-z_j}(t_j, t_0)}{e_{z_j}(t_j, t_0) + e_{-z_j}(t_j, t_0)} \right), \quad t \in [t_0, T],$$

where c_1 is an arbitrary real constant.

Problem 4.8 Find a ChNN trial solution to the BVP in Problem 2.6.

Answer 4.13

$$y_a(t, p) = \frac{B}{h_1(\sigma^2(T), t_0)} h_1(t, t_0) + \frac{A}{h_1(t_0, \sigma^2(T))} h_1(t, \sigma^2(T))$$

$$+h_3(t, t_0)h_3(t, \sigma^2(T)) \left(\sum_{j=1}^{n} v_j \frac{e_{z_j}(t_j, t_0) - e_{-z_j}(t_j, t_0)}{e_{z_j}(t_j, t_0) + e_{-z_j}(t_j, t_0)} \right), \quad t \in [t_0, T],$$

where c_1 is an arbitrary real constant.

Problem 4.9 Find a ChNN trial solution to the BVP in Problem 2.7.

Answer 4.14

$$y_a(t, p) = -c_4 \left(\frac{h_3(t_0, \sigma(T)) - h_2(t_0, \sigma(T))h_2(\sigma(T), t_0)}{(h_1(t_0, \sigma(T)) + h_2(\sigma(T), t_0))(1 - h_1(t_0, \sigma(T)))} + h_3(t_0, \sigma(T)) \right)$$

$$+c_4 \frac{1}{h_2(\sigma(T), t_0)}$$

$$\times \left(\frac{h_3(t_0, \sigma(T)) - h_2(t_0, \sigma(T))h_2(\sigma(T), t_0)}{(h_1(t_0, \sigma(T)) + h_2(\sigma(T), t_0))(1 - h_1(t_0, \sigma(T)))} + h_3(t_0, \sigma(T)) \right) h_2(t, t_0)$$

$$+c_4 \frac{h_3(t_0, \sigma(T)) - h_2(t_0, \sigma(T))h_2(\sigma(T), t_0)}{(h_1(t_0, \sigma(T)) + h_2(\sigma(T), t_0))(1 - h_1(t_0, \sigma(T)))} \frac{h_2(t, \sigma(T))}{h_2(t_0, \sigma(T))}$$

$$+c_4 h_3(t, \sigma(T))$$

$$+h_3(t, t_0)h_3(t, \sigma(T)) \left(\sum_{j=1}^{n} v_j \frac{e_{z_j}(t_j, t_0) - e_{-z_j}(t_j, t_0)}{e_{z_j}(t_j, t_0) + e_{-z_j}(t_j, t_0)} \right), \quad t \in [t_0, T],$$

where c_4 is an arbitrary real constant.

Chapter 5
Legendre Neural Networks

In this chapter, the Legendre neural network (LeNN) model has been introduced for solving dynamic equations on arbitrary time scales. The LeNN trial solution of dynamic equations has been obtained by using the LeNN model for single input and single output system. Initial value problems and boundary value problems have been solved.

Suppose that $\mathbb{T}$ is a time scale with forward jump operator and delta differentiation operator σ and Δ, respectively. In addition, assume that $t_0, T \in \mathbb{T}$ are such that $t_0 < T$.

5.1 Legendre Polynomials on Time Scales

In this section, we introduce Legendre Polynomials on arbitrary time scales.

Definition 5.1 The Legendre polynomials are defined as follows

$$L_0(t, t_0) = 1,$$
$$L_1(t, t_0) = h_1(t, t_0),$$
$$L_{n+1}(t, t_0) = \frac{1}{n+1} \left((2n+1)h_1(t, t_0)L_n(t, t_0) - nL_{n-1}(t, t_0) \right), \quad n \in \mathbb{N}.$$

We have

$$L_1(t, t_0) = t - t_0,$$
$$L_2(t, t_0) = \frac{1}{2} \left(3(t - t_0)L_1(t, t_0) - L_0(t, t_0) \right)$$
$$= \frac{1}{2} \left(3(t - t_0)^2 - 1 \right),$$

© The Author(s), under exclusive license to Springer Nature Switzerland AG 2025
S. Georgiev, *Neural Network Methods for Dynamic Equations on Time Scales*,
SpringerBriefs in Computational Intelligence,
https://doi.org/10.1007/978-3-031-85056-1_5

$$L_3(t, t_0) = \frac{1}{3}\left(5(t - t_0)L_2(t, t_0) - 2L_1(t, t_0)\right)$$

$$= \frac{1}{3}\left(5(t - t_0)\frac{1}{2}\left(3(t - t_0)^2 - 1\right) - 2(t - t_0)\right)$$

$$= \frac{1}{3}\left(\frac{15}{2}(t - t_0)^3 - \frac{5}{2}(t - t_0) - 2(t - t_0)\right)$$

$$= \frac{1}{3}\left(\frac{15}{2}(t - t_0)^3 - \frac{9}{2}(t - t_0)\right)$$

$$= \frac{1}{2}\left(5(t - t_0)^3 - 3(t - t_0)\right),$$

$$L_4(t, t_0) = \frac{1}{4}\left(7(t - t_0)L_3(t, t_0) - 3L_2(t, t_0)\right)$$

$$= \frac{1}{4}\left(7(t - t_0)\frac{1}{3}\left(\frac{15}{2}(t - t_0)^3 - \frac{9}{2}(t - t_0)\right)\right.$$

$$\left. - \frac{3}{2}\left(3(t - t_0)^2 - 1\right)\right)$$

$$= \frac{1}{4}\left(\frac{105}{6}(t - t_0)^4 - \frac{63}{6}(t - t_0)^2 - \frac{9}{2}(t - t_0)^2 + \frac{3}{2}\right)$$

$$= \frac{1}{4}\left(\frac{105}{6}(t - t_0)^4 - 15(t - t_0)^2 + \frac{3}{2}\right)$$

$$= \frac{1}{8}\left(35(t - t_0)^4 - 30(t - t_0)^2 + 3\right), \quad t \in \mathbb{T},$$

and so on.

5.2 Algorithm of Legendre Neural Network Model

In this section, we will describe the algorithm for Legendre neural networks (LeNN).
Consider the following patterns as

$$(t_1, y_1), \quad (t_2, y_2), \quad \ldots, (t_n, y_n),$$

where $t_j \in \mathbb{T}$, $y_j \in \mathbb{R}$, $j \in \{1, \ldots, n\}$, the initial weights $w_{jk} \in \mathbb{R}$, $j \in \{1, \ldots, n\}$, $k \in \{0, 1, \ldots, m\}$, the hidden weights $v_{jk} \in \mathbb{R}$, $j \in \{1, \ldots, n\}$, $k \in \{0, 1, \ldots, m\}$.
Compute

$$z_j = \sum_{k=1}^{m} w_{jk} L_{k-1}(t_j, t_0), \quad j \in \{1, \ldots, n\},$$

and

$$\tanh_{z_j}(t_j, t_0) = \frac{e_{z_j}(t_j, t_0) - e_{-z_j}(t_j, t_0)}{e_{z_j}(t_j, t_0) + e_{-z_j}(t_j, t_0)}, \quad j \in \{1, \dots, n\}.$$

Then, the output of the network is given by the relation

$$y^1(t_j) = v_j \tanh_{z_j}(t_j, t_0), \quad j \in \{1, \dots, n\}.$$

The error is given by the expression

$$E = \sum_{j=1}^{n} \frac{1}{2}(y_j - y^1(t_j))^2.$$

If the error is large, we update the network data as follows

$$w^{1jk} = w_{jk} + \eta(y_j - y^1(t_j))t_j,$$
$$v^{1jk} = v_{jk} + \eta(y_j - y^1(t_j))t_j, \quad j \in \{1, \dots, n\}, \quad k \in \{0, 1, \dots, m\}.$$

We continue this process while we minimize the error.

Example 5.1 Let $\mathbb{T} = \mathbb{Z}$ and the input, initial weights, hidden weight, output and network rate be as in Example 4.1. The neural results are given in Fig. 5.1.

Column1	Column2	Column3	Column4	Column5	Column6	Column7	Column8
t	y_1	w_0	w_1	w_2	v		
2	7	7	3	4	6		
z(0)	z(1)		y^1	E^1			
5	14		1.605634	14.54959			
w_0^1	w_1^1	w_2^1	v^1				
7.058198	3.058198	4.058198	6.010789				
z^1(0)	z^1(1)		y^2	E^2			
5.029099	14.1746		1.596855	14.59699			
w_0^2	w_1^2	w_2^2	v^2				
7.069005	3.069005	4.069005	6.021595				
z^2(0)	z^2(1)		y^3	E^3			
5.034502	14.20701		1.597576	14.59309			

Fig. 5.1 Two network data updates and the corresponding errors for Example 5.1

Column1	Column2	Column3	Column4	Column5	Column6	Column7	Column8
t	y_1	w_0	w_1	w_2	v		
4	15	9	1	5	8		
z(0)	z(1)		y^1	E^1			
6.5	15		1.489796	91.26281			
w_0^1	w_1^1	w_2^1	v^1				
9.730102	1.730102	5.730102	8.054041				
z^1(0)	z^1(1)		y^2	E^2			
6.865051	17.19031		1.401518	92.45936			
w_0^2	w_1^2	w_2^2	v^2				
9.784496	1.784496	5.784496	8.108435				
z^2(0)	z^2(1)		y^3	E^3			
6.892248	17.35349		1.404212	92.42272			

Fig. 5.2 Two network data updates and the corresponding errors for Example 5.2

Example 5.2 Let $\mathbb{T} = 2^{\mathbb{N}_0}$ and the input, output, initial weights, hidden weight and network rate be as in Example 4.2. The neural results are given in Fig. 5.2.

5.3 Learning Algorithm of LeNN Model

The LeNN trial solution $y_a(t, p)$ for dynamic equations on arbitrary time scales with network parameters p (weights, biases) is given in the following manner

$$y_a(t, p) = A(t) + F(t, N(t, p)), \quad t \in \mathbb{T},$$

where $A(t)$ does not contain adjustable parameters and satisfies only the initial and/or boundary conditions and the second term $F(t, N(t, p))$ contains the single output $N(t, p)$ of LeNN with input t and adjustable parameters p.

Consider a three-layer network with one input node t, $m + 1$ initial weights w_j, $j \in \{0, 1, \ldots, m\}$, and one hidden layer consisting of $m + 1$ number of nodes v_j, $j \in \{0, 1, \ldots, m\}$. Then the output is defined in the following manner

$$N(t, p) = \sum_{j=1}^{n} v_j \tanh_{z_j}(t_j, t_0)$$

$$= \sum_{j=1}^{n} v_j \frac{e_{z_j}(t_j, t_0) - e_{-z_j}(t_j, t_0)}{e_{z_j}(t_j, t_0) + e_{-z_j}(t_j, t_0)},$$

where

$$z_j = \sum_{k=1}^{m} w_{jk} L_{k-1}(t_j, t_0), \quad j \in \{1, \ldots, n\},$$

and $L_l(\cdot, t_0)$, $l \in \{0, 1, \ldots, m\}$ are the Legendre polynomials. Training the neural network means updating the parameters so that the error values converge to the desired accuracy.

We have that

$$N_t^{\Delta}(t, p) = \left(\sum_{j=1}^{n} v_j \frac{e_{z_j}(t_j, t_0) - e_{-z_j}(t_j, t_0)}{e_{z_j}(t_j, t_0) + e_{-z_j}(t_j, t_0)} \right)^{\Delta}(t)$$

$$= \sum_{j=1}^{n} v_j \left(\frac{e_{z_j}(t_j, t_0) - e_{-z_j}(t_j, t_0)}{e_{z_j}(t_j, t_0) + e_{-z_j}(t_j, t_0)} \right)^{\Delta}(t)$$

$$= \sum_{j=1}^{n} v_j \left(\frac{1}{(e_{z_j}(t_j, t_0) + e_{-z_j}(t_j, t_0))(e_{z_j}(\sigma(t_j), t_0) + e_{-z_j}(\sigma(t_j), t_0))} \right.$$

$$\left((z_j e_{z_j}(t_j, t_0) + z_j e_{-z_j}(t_j, t_0))(e_{z_j}(t_j, t_0) + e_{-z_j}(t_j, t_0)) \right.$$

$$\left. \left. - (e_{z_j}(t_j, t_0) - e_{-z_j}(t_j, t_0))(z_j e_{z_j}(t_j, t_0) - z_j e_{-z_j}(t_j, t_0)) \right) \right)$$

$$= \sum_{j=1}^{n} v_j z_j \left(\frac{1}{(e_{z_j}(t_j, t_0) + e_{-z_j}(t_j, t_0))(e_{z_j}(\sigma(t_j), t_0) + e_{-z_j}(\sigma(t_j), t_0))} \right.$$

$$\left. \left((e_{z_j}(t_j, t_0) + e_{-z_j}(t_j, t_0))^2 - (e_{z_j}(t_j, t_0) - e_{-z_j}(t_j, t_0))^2 \right) \right)$$

$$= \sum_{j=1}^{n} v_j z_j \frac{4 e_{z_j}(t_j, t_0) e_{-z_j}(t_j, t_0)}{(e_{z_j}(t_j, t_0) + e_{-z_j}(t_j, t_0))(e_{z_j}(\sigma(t_j), t_0) + e_{-z_j}(\sigma(t_j), t_0))}$$

$$= \sum_{j=1}^{n} v_j z_j \frac{4 e_{z_j \oplus (-z_j)}(t_j, t_0)}{(e_{z_j}(t_j, t_0) + e_{-z_j}(t_j, t_0))(e_{z_j}(\sigma(t_j), t_0) + e_{-z_j}(\sigma(t_j), t_0))}.$$

5.4 Formulation for Initial Value Problems

In this section, we will use some computations from Chap. 2.

1. Consider the IVP (2.2). Then the LeNN trial solution is given by the expression

$$y_a(t, p) = y_0 + h_1(t, t_0) \left(\sum_{j=1}^{n} v_j \frac{e_{z_j}(t_j, t_0) - e_{-z_j}(t_j, t_0)}{e_{z_j}(t_j, t_0) + e_{-z_j}(t_j, t_0)} \right), \quad t \in [t_0, T].$$

2. Consider the IVP (2.4). Then the LeNN trial solution is given by the expression

$$y_a(t, p) = y_0 + y_1 h_1(t, t_0) + h_2(t, t_0) \left(\sum_{j=1}^{n} v_j \frac{e_{z_j}(t_j, t_0) - e_{-z_j}(t_j, t_0)}{e_{z_j}(t_j, t_0) + e_{-z_j}(t_j, t_0)} \right), \quad t \in [t_0, T].$$

3. Consider the IVP (2.5). Then the LeNN trial solution is given by the expression

$$y_a(t, p) = \sum_{j=0}^{n-1} y_j h_j(t, t_0) + h_n(t, t_0) \left(\sum_{j=1}^{n} v_j \frac{e_{z_j}(t_j, t_0) - e_{-z_j}(t_j, t_0)}{e_{z_j}(t_j, t_0) + e_{-z_j}(t_j, t_0)} \right), \quad t \in [t_0, T].$$

5.5 Formulation for Boundary Value Problems

In this section, we will use some computations in Chap. 2.

1. Consider the BVP (2.7). Then the LeNN trial solution is given by the expression

$$y_a(t, p) = C + \frac{D - C}{h_1(T, t_0)} h_1(t, t_0)$$

$$+ h_2(t, t_0) h_2(t, T) \left(\sum_{j=1}^{n} v_j \frac{e_{z_j}(t_j, t_0) - e_{-z_j}(t_j, t_0)}{e_{z_j}(t_j, t_0) + e_{-z_j}(t_j, t_0)} \right), \quad t \in [t_0, T].$$

2. Consider the BVP (2.8). Then the LeNN trial solution is given by the expression

$$y_a(t, p) = c_1 - \frac{(M + R)c_1}{R h_1(\sigma(T), t_0)} h_1(t, t_0) + h_2(t, t_0) h_2(t, \sigma(T))$$

$$\times \left(\sum_{j=1}^{n} v_j \frac{e_{z_j}(t_j, t_0) - e_{-z_j}(t_j, t_0)}{e_{z_j}(t_j, t_0) + e_{-z_j}(t_j, t_0)} \right), \quad t \in [t_0, T],$$

where c_1 is an arbitrary real constant.

Exercise 5.1 Find the LeNN trial solution for the BVP in Exercise 2.1.

Answer 5.1

$$y_a(t, p) = c + h_2(t, t_0)h_2(t, \sigma(T)) \left(\sum_{j=1}^{n} v_j \frac{e_{z_j}(t_j, t_0) - e_{-z_j}(t_j, t_0)}{e_{z_j}(t_j, t_0) + e_{-z_j}(t_j, t_0)} \right), \quad t \in [t_0, T],$$

where c is an arbitrary real constant.

3. Consider the BVP (2.9). Then the LeNN trial solution is given by the expression

$$y_a(t, p) = C + Dh_1(t, t_0) + h_2(t, t_0)h_2(t, T) \left(\sum_{j=1}^{n} v_j \frac{e_{z_j}(t_j, t_0) - e_{-z_j}(t_j, t_0)}{e_{z_j}(t_j, t_0) + e_{-z_j}(t_j, t_0)} \right),$$

$$t \in [t_0, T].$$

4. Consider the BVP (2.10). Then the LeNN trial solution is given by the expression

$$y_a(t, p) = \frac{\beta}{\alpha}c_2 + c_2 h_1(t, t_0) - c_2 \frac{\gamma\beta + \alpha\gamma h_1(\sigma^2(t_0), t_0) + \alpha\delta}{\alpha(\gamma h_2(\sigma^2(t_0), t_0) + \delta h_1(\sigma(T), t_0))} h_2(t, t_0)$$

$$+ h_2(t, t_0)h_1(t, \sigma^2(t_0))h_2(t, \sigma(T)) \left(\sum_{j=1}^{n} v_j \frac{e_{z_j}(t_j, t_0) - e_{-z_j}(t_j, t_0)}{e_{z_j}(t_j, t_0) + e_{-z_j}(t_j, t_0)} \right),$$

$$t \in [t_0, T],$$

where c_2 is an arbitrary real constant.

Exercise 5.2 Find the LeNN trial solution for the BVP in Exercise 2.2

Answer 5.2

$$y_a(t, p) = c_2 h_1(t, t_0) - c_2 \frac{1}{h_1(\sigma(T), t_0)} h_2(t, t_0) + h_2(t, t_0)h_1(t, \sigma^2(t_0))h_2(t, \sigma(T))$$

$$\times \left(\sum_{j=1}^{n} v_j \frac{e_{z_j}(t_j, t_0) - e_{-z_j}(t_j, t_0)}{e_{z_j}(t_j, t_0) + e_{-z_j}(t_j, t_0)} \right), \quad t \in [t_0, T],$$

where c_2 is an arbitrary real constant.

5. Consider the BVP (2.16). Then the LeNN trial solution is given by the expression

$$y_a(t, p) = \sum_{j=0}^{n-2} c_j h_j(t, t_0) + h_n(t, t_0)h_n(t, \sigma(T)) \left(\sum_{j=1}^{n} v_j \frac{e_{z_j}(t_j, t_0) - e_{-z_j}(t_j, t_0)}{e_{z_j}(t_j, t_0) + e_{-z_j}(t_j, t_0)} \right),$$

$$t \in [t_0, T].$$

Column1	Column2	Column3	Column4	Column5	Column6	Column7	Column8
t	y_1	w_0	w_1	w_2	v		
2	0.333333	7	3	4	6		
z(0)	z(1)		y^1	E^1			
5	14		1.605634	0.809374			
w_0^1	w_1^1	w_2^1	v^1				
7.003237	3.003237	4.003237	5.997455				
z^1(0)	z^1(1)		y^2	E^2			
5.001619	14.00971		1.6043	0.807678			
w_0^2	w_1^2	w_2^2	v^2				
7.000696	3.000696	4.000696	5.994913				
z^2(0)	z^2(1)		y^3	E^3			
5.000348	14.00209		1.604132	0.807465			

Fig. 5.3 Two network data updates and the corresponding errors for Example 5.3

5.6   Examples

The results in this chapter we will illustrate with the following examples.

Example 5.3 Let $\mathbb{T} = \mathbb{Z}$. Consider the IVP, input, output, initial weights, hidden weight and network rate as in Example 4.3. The neural results are given in Fig. 5.3.

Example 5.4 Let $t = 2^{\mathbb{N}_0}$. Consider the BVP, input, output, initial weights, hidden weight and network rate as in Example 4.4. The neural results are given in Fig. 5.4.

5.7   Advanced Practical Problems

Problem 5.1 Find a LeNN trial solution for the m point BVP in Problem 2.1.

Answer 5.3

$$y_a(t, p) = \sum_{k=1}^{m-1} c_k h_1(t, t_k) + c_m h_1(t, \sigma(t_m)) + \prod_{k=1}^{m-1} h_2(t, t_k) h_2(t, \sigma(t_m))$$

$$\times \left(\sum_{j=1}^{n} v_j \frac{e_{z_j}(t_j, t_0) - e_{-z_j}(t_j, t_0)}{e_{z_j}(t_j, t_0) + e_{-z_j}(t_j, t_0)} \right), \quad t \in [t_0, T],$$

Column1	Column2	Column3	Column4	Column5	Column6	Column7	Column8
t	y_1	w_0	w_1	w_2	v		
4	0.125	9	1	5	8		
z(0)	z(1)		y^1	E^1			
6.5	15		1.489796	0.931334			
w_0^1	w_1^1	w_2^1	v^1				
9.007451	1.007451	5.007451	7.994541				
z^1(0)	z^1(1)		y^2	E^2			
6.503725	15.02235		1.4877	0.928475			
w_0^2	w_1^2	w_2^2	v^2				
9.002	1.002	5.002	7.98909				
z^2(0)	z^2(1)		y^3	E^3			
6.501	15.006		1.487474	0.928168			

Fig. 5.4 Two network data updates and the corresponding errors for Example 5.4

where $c_1, \ldots, c_m$ are real constants that satisfy the system

$$\sum_{k=2}^{m-1} c_k h_1(t_1, t_k) + c_m h_1(t_1, \sigma(t_m)) = \sum_{k=2}^{m-1} \alpha_k \left(\sum_{l=1, l \neq k}^{m-1} c_l h_1(t_l, t_k) + c_m h_1(t_k, \sigma(t_m)) \right)$$
$$+ \alpha_1 \sum_{k=1}^{m-1} c_k h_1(\sigma(t_m), t_k).$$

Problem 5.2 Find a LeNN trial solution to the BVP in Problem 2.2.

Answer 5.4

$$y_a(t, p) = c_2 + c_2 h_1(t, t_0) - c_2 \frac{1}{h_1(\sigma(T), t_0)} h_2(t, t_0) + h_2(t, t_0) h_1(t, \sigma^2(t_0)) h_2(t, \sigma(T))$$
$$\times \left(\sum_{j=1}^{n} v_j \frac{e_{z_j}(t_j, t_0) - e_{-z_j}(t_j, t_0)}{e_{z_j}(t_j, t_0) + e_{-z_j}(t_j, t_0)} \right), \quad t \in [t_0, T],$$

where c_2 is an arbitrary real constant.

Problem 5.3 Find a LeNN trial solution to the BVP in Problem 2.3.

Answer 5.5

$$y_a(t, p) = c_1 - \frac{c_1}{h_2(\sigma^2(T), t_0)} h_1(t, t_0) - \frac{c_1}{h_2(t_0, \sigma^2(T))} h_2(t, \sigma^2(T))$$

$$+h_3(t, t_0)h_3(t, \sigma^2(T)) \left(\sum_{j=1}^{n} v_j \frac{e_{z_j}(t_j, t_0) - e_{-z_j}(t_j, t_0)}{e_{z_j}(t_j, t_0) + e_{-z_j}(t_j, t_0)} \right), \quad t \in [t_0, T],$$

where c_1 is an arbitrary real constant.

Problem 5.4 Find a LeNN trial solution to the BVP in Problem 2.4.

Answer 5.6

$$y_a(t, p) = c_1 - \frac{c_1}{h_2(\sigma(T), t_0)} h_1(t, t_0) - \frac{c_1}{h_2(t_0, \sigma(T))} h_2(t, \sigma(T))$$

$$+h_3(t, t_0)h_3(t, \sigma(T)) \left(\sum_{j=1}^{n} v_j \frac{e_{z_j}(t_j, t_0) - e_{-z_j}(t_j, t_0)}{e_{z_j}(t_j, t_0) + e_{-z_j}(t_j, t_0)} \right), \quad t \in [t_0, T],$$

where c_1 is an arbitrary real constant.

Problem 5.5 Find a LeNN trial solution to the BVP in Problem 2.5.

Answer 5.7

$$y_a(t, p) = c_1 h_1(t, t_0) - c_1 h_1(t, \eta) + \frac{(k-1)c_1 h_1(\eta, t_0)}{h_2(t_0, \sigma(T)) - kh_2(\eta, \sigma(T))} h_2(t, \sigma(T))$$

$$+h_2(t, t_0)h_2(t, \eta)h_3(t, \sigma(T)) \left(\sum_{j=1}^{n} v_j \frac{e_{z_j}(t_j, t_0) - e_{-z_j}(t_j, t_0)}{e_{z_j}(t_j, t_0) + e_{-z_j}(t_j, t_0)} \right), \quad t \in [t_0, T],$$

where c_1 is an arbitrary real constant.

Problem 5.6 Find a LeNN trial solution to the BVP in Problem 2.6.

Answer 5.8

$$y_a(t, p) = \frac{B}{h_1(\sigma^2(T), t_0)} h_1(t, t_0) + \frac{A}{h_1(t_0, \sigma^2(T))} h_1(t, \sigma^2(T))$$

$$+h_3(t, t_0)h_3(t, \sigma^2(T)) \left(\sum_{j=1}^{n} v_j \frac{e_{z_j}(t_j, t_0) - e_{-z_j}(t_j, t_0)}{e_{z_j}(t_j, t_0) + e_{-z_j}(t_j, t_0)} \right), \quad t \in [t_0, T],$$

where c_1 is an arbitrary real constant.

Problem 5.7 Find a LeNN trial solution to the BVP in Problem 2.7.

Answer 5.9

$$y_a(t, p) = -c_4 \left(\frac{h_3(t_0, \sigma(T)) - h_2(t_0, \sigma(T))h_2(\sigma(T), t_0)}{(h_1(t_0, \sigma(T)) + h_2(\sigma(T), t_0))(1 - h_1(t_0, \sigma(T)))} + h_3(t_0, \sigma(T)) \right)$$

$$+c_4 \frac{1}{h_2(\sigma(T), t_0)}$$

$$\times \left(\frac{h_3(t_0, \sigma(T)) - h_2(t_0, \sigma(T)) h_2(\sigma(T), t_0)}{(h_1(t_0, \sigma(T)) + h_2(\sigma(T), t_0))(1 - h_1(t_0, \sigma(T)))} + h_3(t_0, \sigma(T)) \right) h_2(t, t_0)$$

$$+c_4 \frac{h_3(t_0, \sigma(T)) - h_2(t_0, \sigma(T)) h_2(\sigma(T), t_0)}{(h_1(t_0, \sigma(T)) + h_2(\sigma(T), t_0))(1 - h_1(t_0, \sigma(T)))} \frac{h_2(t, \sigma(T))}{h_2(t_0, \sigma(T))}$$

$$+c_4 h_3(t, \sigma(T))$$

$$+h_3(t, t_0) h_3(t, \sigma(T)) \left(\sum_{j=1}^{n} v_j \frac{e_{z_j}(t_j, t_0) - e_{-z_j}(t_j, t_0)}{e_{z_j}(t_j, t_0) + e_{-z_j}(t_j, t_0)} \right), \quad t \in [t_0, T],$$

where c_4 is an arbitrary real constant.

Index

© The Editor(s) (if applicable) and The Author(s), under exclusive license
to Springer Nature Switzerland AG 2025
S. Georgiev, *Neural Network Methods for Dynamic Equations on Time Scales*,
SpringerBriefs in Computational Intelligence,
https://doi.org/10.1007/978-3-031-85056-1